Computer Games as a Sociocultural Phenomenon

Computer Games as a Sociocultural Phenomenon

Games Without Frontiers
War Without Tears

Edited by

Andreas Jahn-Sudmann and Ralf Stockmann
Georg-August University of Göttingen

Selection and editorial matter © Andreas Jahn-Sudmann and
Ralf Stockmann 2008
Chapters © their authors 2008

First published 2008 by
PALGRAVE MACMILLAN
Houndmills, Basingstoke, Hampshire RG21 6XS and
175 Fifth Avenue, New York, N.Y. 10010
Companies and representatives throughout the world

PALGRAVE MACMILLAN is the global academic imprint of the Palgrave
Macmillan division of St. Martin's Press, LLC and of Palgrave Macmillan Ltd.
Macmillan® is a registered trademark in the United States, United Kingdom
and other countries. Palgrave is a registered trademark in the European
Union and other countries.

ISBN-13: 978-0-230-54544-1   hardback
ISBN-10: 0-230-54544-0   hardback

This book is printed on paper suitable for recycling and made from fully
managed and sustained forest sources. Logging, pulping and manufacturing
processes are expected to conform to the environmental regulations of the
country of origin.

A catalogue record for this book is available from the British Library.

A catalog record for this book is available from the Library of Congress.

10   9   8   7   6   5   4   3   2   1
17   16   15   14   13   12   11   10   09   08

Printed and bound in Great Britain by
CPI Antony Rowe, Chippenham and Eastbourne

Contents

Acknowledgements viii

Notes on the Contributors ix

Introduction by Andreas Jahn-Sudmann and Ralf Stockmann xiii

Part I: Game Design and Aesthetics

1   The Aesthetic Vocabulary of Video Games 3
    Joost van Dreunen

2   Can Games Get Real? A Closer Look at 'Documentary'
    Digital Games 12
    Ian Bogost and Cindy Poremba

3   Emotional Design of Computer Games and Fiction Films 22
    Doris C. Rusch

4   'Applied Game Theory': Innovation, Diversity,
    Experimentation in Contemporary Game Design 32
    Henry Jenkins and Kurt Squire

Part II: Space and Time

5   There and Back Again: Reuse, Signifiers and Consistency
    in Created Game Spaces 47
    Peter Berger

6   Another Bricolage in the Wall: Deleuze and
    Teenage Alienation 56
    Jeffrey P. Cain

Part III: War and Violence

7   Programming Violence: Language and the Making of
    Interactive Media 69
    Claudia Herbst

8   Impotence and Agency: Computer Games as
    a Post-9/11 Battlefield 78
    Henry Lowood

v

9   S(t)imulating War: From Early Films to Military Games   87
    Daphnée Rentfrow

Part IV: Ethics and Morality

10   Player in Fabula: Ethics of Interaction as Semiotic
     Negotiation Between Authorship and Readership   99
     Massimo Maietti

11   'Moral Management': Dealing with Moral Concerns to
     Maintain Enjoyment of Violent Video Games   108
     Christoph Klimmt, Hannah Schmid, Andreas Nosper,
     Tilo Hartmann and Peter Vorderer

12   Beyond Good and Evil: The Inhuman Ethics of
     Redemption and *Bloodlines*   119
     Will Slocombe

Part V: Politics and Ideology

13   Preconscious Apocalypse: The Failure of Capitalism in
     Computer Games   131
     Sven O. Cavalcanti

14   Borders and Bodies in *City of Heroes*: (Re)imaging
     American Identity Post 9/11   140
     Nowell Marshall

15   Anti-PC Games: Exploring Articulations of the
     Politically Incorrect in *GTA San Andreas*   150
     Andreas Jahn-Sudmann and Ralf Stockmann

16   Strip – Shift – Impose – Recycle – Overload – Spill –
     Breakout – Abuse. Artists' (Mis-)Appropriations of
     Shooter Games   162
     Maia Engeli

Part VI: Computer Game Play(ers) and Cultural Identities

17   Presence-Play: The Hauntology of the Computer Game   175
     Dean Lockwood and Tony Richards

18   Negotiating Online Computer Games in East Asia:
     Manufacturing Asian MMORPGs and Marketing
     'Asianness'   186
     Dean Chan

19 Teenage Girls 'Play House': The Cyber-drama of
    The Sims   197
    Lynda Dyson

Cited Computer Games   207

References   211

Index   226

Acknowledgements

Many thanks are due to the contributors, who treated with admirable patience the prolonged production process of this collection. We thank them all for their hard work in producing a volume that illustrates the rich diversity of digital games culture and international game studies.

Very sincere thanks are due to Jill Lake, Melanie Blair, Tim Kapp, Christabel Scaife and Rick Bouwman at Palgrave Macmillan for their help and support during the birth process of the book.

The editors would also like to thank Robin Oppenheimer, Nowell Marshall, Henry Lowood and Dean Lockwood for proofreading and their invaluable assistance.

The subtitle is taken from Peter Gabriel's song 'Games Without Frontiers' (1980).

Finally, we are very grateful to Kirsten Jahn for her constant support. Our debt to her is immeasurable.

Notes on the Contributors

Peter Berger is a software developer for Apple. He also reviews and critiques games and game design for *Played To Death* magazine (http://played.todeath.com) and *Tea Leaves* (http://tleaves.com).

Ian Bogost is Assistant Professor of Digital Media at the Georgia Institute of Technology, USA and Founding Partner at Persuasive Games LLC. He is author of *Unit Operations: An Approach to Videogame Criticism* (2006) and *Persuasive Games: The Expressive Power of Videogames* (2007).

Jeffrey P. Cain is Associate Professor of English at Sacred Heart University in Fairfield, Connecticut, USA. His research interests include poststructuralist literary theory, early-modern literature and philosophy, and popular and rural culture.

Sven O. Cavalcanti is a PhD candidate at the Institute for Sociology and Social Psychology at Hanover University, Germany. He has published numerous articles on social theory, politics and culture.

Dean Chan is a Postdoctoral Research Fellow at the School of Communications and Contemporary Arts, Edith Cowan University, in Perth, Western Australia. His current research interests focus on East Asian console and multiplayer online games, diasporic Asian gamers, racialised representational politics in videogames, and digital game art.

Joost van Dreunen is a PhD candidate at Columbia University, New York, USA as well as a project director at the Columbia Institute for Tele-Information and a member of the Center for Organizational Innovation. His research focuses on the cultural and epistemological implications of video games.

Lynda Dyson is Senior Lecturer in Media and Cultural Studies at London College of Communication, University of the Arts, London, UK.

Maia Engeli is Assistant Professor at the School of Interactive Arts and Technology at Simon Fraser University, BC, Canada and works in the area of telematic architectures for collaborative productive processes, learning, and entertainment. She teaches interactive media design, immersive environments and computer game modding and is author of *bits and spaces* (2001) and *Digital Stories – The Poetics of Communication*

(2000). Until 2002 she was Assistant Professor and acting head of the chair for Architecture and CAAD at the Swiss Federal Institute of Technology (ETH) in Zurich, Switzerland.

Tilo Hartmann is a postdoctoral research assistant at the Institute of Mass Communication and Media Research, University of Zurich, Switzerland. Previous affiliations include positions at the Hanover University of Music and Drama (2001-2005), University of Southern California (2006), and University of Erfurt (2006). His research focuses on media use (selective exposure, experience, effects).

Claudia Herbst is Associate Professor in the Department of Digital Art at Pratt Institute, Brooklyn, New York, USA. She is a regular contributor to anthologies and journals; in her writing, Herbst investigates questions of language and technology.

Henry Jenkins is the Deflorz Professor of Humanities and the Co-director of the Comparative Media Studies Program at MIT. His most recent books include *Convergence Culture: Where Old and New Media Collide, Fans, Bloggers, and Gamers: Exploring Participatory Culture*, and *The Wow Climax: Emotion in Popular Culture*.

Christoph Klimmt is a postdoctoral research assistant at Hanover University of Music and Drama, Department of Journalism and Communication Research (IJK), Germany. Since 2006 he has been team leader in the European research project 'The Fun of Gaming' (FUGA), temporary occupation of a professorship in media science at IJK, and co-leader of a research project on the effects of violent video games funded by the German Research Foundation.

Dean Lockwood is a Senior Lecturer in Media Theory at the University of Lincoln. He has recently published on the 'spectacle of the real' and is currently researching various kinds of deconstructive monstrosity in the fields of digital culture and horror.

Henry Lowood is curator of History of Science and Technology Collections at Stanford University, USA, and a historian of technology. He has led Stanford's Silicon Valley Archives since its inception. He is also co-principal investigator of 'How They Got Game: The History and Culture of Interactive Simulations and Videogames', a project funded by the Stanford Humanities Laboratory. This project has dedicated much of its effort to historical studies of military simulation and the growth of the 'military-entertainment complex'. His Stanford course, 'The History of Computer Game Design: Business, Culture, Technology', is one of the first devoted to the history of this medium.

Massimo Maietti is a researcher in the field of semiotics of interactive media. He is the author of *Semiotica dei Videogiochi* (*Semiotics of Videogames*, 2004). He has lectured at various universities in Italy, the United Kingdom and Spain. He works in London as a game designer.

Nowell Marshall is a doctoral candidate studying gothic and Romantic literature and theories of the corporeal-affective body at the University of California, Riverside, USA. His dissertation traces the roots of depression in gender-variant people through the literature and cultural history of the British Romantic period.

Andreas Nosper is project manager at the media and marketing consulting agency aserto in Hanover, Germany. His research focuses on media entertainment, user-generated media content and corporate communication.

Cindy Poremba is a digital-media theorist, producer and curator researching documentary and videogames through the interdisciplinary Humanities Doctoral program at Concordia University, Montreal, Canada. Her work focuses on rhetoric, feminist and documentary theory as it intersects with cultural memory, recombinant poetics, creative constructionism and other forms of digital practice, particularly in the context of games and robotics.

Daphnée Rentfrow received her PhD in Comparative Literature from Brown University where she became a member of the research faculty, serving as manager for the digital archive The Modernist Journals Project. While her primary field of study is literature and culture of the First World War, she is also a scholar of digital humanities and has published on thematic research collections and women's studies, the role of techno-pedagogy in today's university, and contemporary concerns in librarianship, digital collection management and digital collections in American Studies.

Tony Richards teaches video and new-media theory on the media production degree at the University of Lincoln. His research interests mainly revolve around deconstruction and its relation to new media, especially video games. He is currently also working on an interactive database linking media theory to student practice on the degree.

Doris C. Rusch is an affiliated researcher with the MIT Convergence Culture Consortium and lecturer in critical game studies at the Danube University Krems. She has done postdoctoral work at the Institute for Design and Assessment of Technology at Vienna University of Technology. Her research focuses on the medium-specific potentials of digital games and how they can be employed to design thought-provoking, meaning-ful and emotionally rich experiences.

Hannah Schmid is research executive at the market research company Synovate in Munich, Germany. Her research focuses on media effects and media entertainment, particularly video games and global entertainment.

Will Slocombe is Lecturer in Twentieth-Century Literature at the University of Wales, Aberystwyth, UK, and teaches American and British literature, literary theory and postmodern fictions, and supervises research on science fiction, technology and new media. He is the author of *Nihilism and the Sublime Postmodern* (2006) and various articles on postwar literature, nihilism and computer games ethics. He also sits on the editorial board of *Writing Technologies*.

Kurt Squire is Assistant Professor of Educational Communications and Technology at the University of Wisconsin-Madison, USA and Director of UW-Madisons's Games, Learning and Society Initiative.

Peter Vorderer is Professor of Communication and Psychology at the Annenberg School for Communication and the Department of Psychology, University of Southern California, Los Angeles, USA. He is editor of *Media Psychology* and former editor of *Zeitschrift für Medienpsychologie*. His primary research interests are media effects and media entertainment, particularly video games.

Introduction

Andreas Jahn-Sudmann and Ralf Stockmann

Computer games are a sociocultural phenomenon of increasing relevance. They have left behind their early stage as a 'youth and children's medium' and are now being used by broad levels of society as essential recreational activities, and thus are of considerable economic importance. In 2005, the game industry's software sales were considerably higher, $30 billion, than Hollywood's revenues from worldwide theatrical film releases that, according to the Motion Picture Association of America, amounted to $23 billion (Müller-Lietzkow, Bouncken and Seufert, 2006). Meanwhile, the most popular Massively Multiplayer Online Role-Playing Game (MMORPG), *World of Warcraft*, (2004) connects eight million paying customers worldwide. And the release of each new Nintendo game console turns out to be a consumer event: in Europe Nintendo's Wii, with its innovative interface, was sold out within a week.

Moreover, the game industry not only makes money out of the production and distribution of games, but out of game-playing itself. Professional gamers compete in e-sport leagues and tournaments. In countries such as Korea, prime gamers can clear about $500,000 a year. On online market places, in-game items from MMORPGs such as *Dark Age of Camelot* (2001) or *City of Heroes* (2004) are sold for large sums and companies have been launched that offer to level up or build up the features and abilities of online role-playing games' characters.

Computer games are omnipresent. You not only play them in the privacy of your own computer or your own game console, but also in public on your cell phone or handheld, on the streets, in swimming pools, while travelling by rail, and not least of all, at the office.

Regardless of their increasing popularity and presence, computer games remain the focus of mainstream media attention when horrific acts of violence (war and rampage – Iraq and Columbine) draw public attention. The first impetus, then, is aimed at a more or less serious examination of their dangerous and problematic aspects (blunting people's senses, playing down and provoking violence, players' loss of contact with reality, escapism, unscrupulousness, and so on).

Littleton Columbine High School, Erfurt Gutenberg Gymnasium and other schools where teenagers shot other teenagers – the fact that the offenders in each case had played first-person shooter games such as *Doom* (1993) or *Quake* (1996) was enough to reduce most subsequent media debates to the simple formula: 'killer games' create 'killers'. Accordingly, it was considered imperative to tighten the legal protection for children and young persons or completely ban those games whose violent content is classified as potentially dangerous.

The hasty and occasionally naïve reactions of some politicians, reporters and advocacy organizations demonstrate that parts of the public (particularly the part that does not know anything about or opposes computer games) automatically expect, when faced with such horrifying news, that politicians should somehow swiftly respond to violent acts.

Prohibiting games suggests itself as a comfortable and popular solution, primarily because the responsibility for violent acts can mainly be put on the media and does not have to be discussed as a profound social problem that refuses simple explanations. Advocates of tighter laws feel their assumptions justified by a number of scientific papers (Anderson, 2004). Nevertheless, the question of to what extent media violence in general, and violent computer games in particular, produce or forward (lasting) aggressive behaviour is still the subject of an extremely controversial discussion. A scientific consensus, as has repeatedly been claimed, does not exist (Newman, 2004, p. 66).

The negative image of computer games, however, is not restricted to the representation, performance and virtual acts of violence and their effects. Computer games are still considered to be a time-consuming occupation that supports no cultural gratification (knowledge, intellectual reflection, basic life skills) except for the mere pleasure of playing. Compared with other forms of media culture (internet, TV, movies, radio), whose products are more likely to be credited with the ability to advance cultural needs, computer games are still regarded as trivial and one-dimensional activities that serve only as diversions. Above all, computer games are supposed to be addictive and accompanied by social isolation. Finally, despite recent empirical findings, playing computer games is still classified as a childish or adolescent activity (Newman, 2004, p. 5).

Apart from the ongoing discussions about the problematic aspects of computer games, there exists a growing awareness of their productive and creative potential. Correspondingly, games are frequently used for learning and training purposes within and outside of the professional sphere (Prensky, 2001), or are used therapeutically (Griffiths, 2005). Also, various (institutional) efforts are made to promote the official approval and acknowledgement of games as cultural assets. 'Museums and art programs have begun to incorporate video games into their exhibits and curricula as games begin to achieve recognition in the art world. Like the great figures we expect to find occupying key places in an artistic canon, there are game designers who have reached *auteur* status' (Smuts, 2005). More and more, game festivals award prizes to 'culturally and aesthetically valuable' computer games; these include the Independent Game Festival in San Francisco that has been awarding innovative independent computer game producers since 1998 or the London Games Festival that has supported the British Video Game Awards.

Last but not least, computer games have undoubtedly become an important subject of academic discourse in recent years. According to Mark J.P. Wolf and Bernard Perron (2003, p. 1), 'the video game has recently become the hottest and most volatile field of study within new media theory'. Henry Jenkins (2002) has called video games 'an art form for the digital age... shaping the aesthetic sensibility of the 21st century'.

Meanwhile, researchers are no longer exclusively interested in digital games' problematic implications or effects, but are analysing them according to aesthetic, narrative, economic and technical aspects. This broadening of perspectives and growing interest in games is the result of, among other things, the fact that the first generation of academics that has grown up with computer games is now applying itself to the topic and its respective disciplinary fields (Neitzel and Rohr, 2006, p. 9).

Currently, international game studies are, according to Jesper Juul (2001a), in a state of productive chaos. It is a comparatively young discipline whose profile still has to evolve. Basic determinations and alignments about approaches, goals and subjects of game studies are vehemently discussed, as can be observed in the controversial debate between narratologists and ludologists. In this book, we will try to present new scientific approaches to basic theoretical and analytical questions of game studies. The complexity of establishing game studies as an academic discipline, despite a growing number of trade journals, websites, study groups and conferences, can be deduced from one constantly repeated postulation – which is, to take games seriously as a subject of scientific research. The ongoing negative image of computer games in which they were for a long time rated only slightly higher than pornography (Chaplin and Ruby, 2005, p. 1) certainly slows down the process of institutionalizing and emphasizing game studies in the academic world.

Of course, the acknowledgement of game studies cannot be reduced to cultural upgrading and avoiding everything else that might reinforce its negative image. The point is not to take games seriously, but to take them seriously in a critical way. This implies keeping an eye on, among other things, the role of violence in games; not only because of the ongoing public debates and calls for the prohibition of violent computer games, but particularly because violence is a part of almost every computer game. Therefore, it is important to thoroughly analyse and critically discuss the various ideological-political, moral, ethical, aesthetic, narrative and cultural contexts and manifestations of violence in games. For this reason we will not exclude the connection between games and violence from this book. Using the example of violent and non-violent computer games, we will try to understand the productive potentials and imaginative possibilities of games and show critical perspectives on games that have so far been neglected in the discussions around digital entertainment.

Structure

This book examines digital games at a number of levels and within various contexts. The texts do not share a common research approach. Our aim is rather to present a broad array of analytical perspectives in the examination of computer games, some of which take and use contradictory positions and concepts: some are rather empirical, while others have a theoretical focus. The authors of this interdisciplinary project are game designers, artists and scientists in various research fields, who inhabit broad and multifaceted views on the multiple forms of articulation and utilization of digital games. This volume contains six thematic parts that represent central discourse fields of present games research.

Part I: Game Design and Aesthetics

The first thematic block assembles articles that address basic theoretical and methodological problems, concepts and models of game design and aesthetics. The focus is to map out the specific qualities of computer games, particularly in comparison to other media, and to present and develop new theoretical approaches in the research of digital games and gaming. Apart from the debate on the exertion and depiction of violence in computer games, this certainly is *the* central discourse in the field of game studies. Henry Jenkins and Kurt Squire analyse the aesthetics and Joost van Dreunen focuses on the aesthetic vocabulary of computer games in particular. Doris C. Rusch delineates a model of the emotional design of computer games, while Ian Bogost and Cindy Poremba present a contemplative model of the 'documentary' digital game as non-fiction game genre.

Part II: Space and Time

At all times, computer games have constituted a geographical space – whether as the restricted playground of *Pong* (1972) or the almost infinite vastness of current MMORPGs such as *World of Warcraft*. Articles in this chapter investigate configurations of time and space in games and their relationship to the game experience. On the one hand, Peter Berger's question relates to which techniques can be applied to increase players' immersion in a created virtual space and improve the mimesis of that space. On the other hand, Jeffrey P. Cain applies the concepts of Gilles Deleuze and Félix Guattari to a cultural analysis of computer games, especially the influential MMORPG *EverQuest* (1999). These concepts provide vantage points from which to consider the physical, narrative and virtual spaces in which video games occur.

Part III: War and Violence

Violent games, particularly war games, are exceedingly popular, among both juveniles and grown-up players. While the discussion about the effects of the use of games with violent content broke out in the 1970s, the interrelationship between war and games became the centre of public and scientific discussion when the images of the second Gulf War in 1990–91 were broadcast. To create the illusion of a clinical operation, TV images were presented like a computer simulation which promised absolute precision and minimum 'collateral damage' that was meant to erase the real war's 'reality' and its consequences. Therefore, more realistic and immediate scenarios of urban battlefields in the war against terror appear at the beginning of the 21st century. Long ago, the US military discovered the value of virtual warfare for its own purposes, as a means of improving its image, recruiting soldiers and employing educational methodologies. The co-operation between the US military and Hollywood has a longlasting tradition, and it now includes the game industry. Henry Lowood describes the situation immediately after the terror attacks of 9/11, when game publishers shelved computer games reminiscent of the events, whereas players used game mods to express their rage. Daphnée Rentfrow turns to early film to map how new technologies aspire to a fidelity between the representation and the reality of war and show that this is not new. And Claudia Herbst shows how code informs violent gaming context, as language, narrative structures and content and inseparably intertwined.

Part IV: Ethics and Morality

To date, researchers have not satisfactorily analysed questions of ethics and morality in computer games. In an unprecedented way, computer games provide the possibility to decide between 'good' and 'evil' game behaviour. In part, this belongs to the game core of principles (*Black & White*, 2001;

Star Wars: Knights of the Old Republic, 2003). Without doubt, the fascination of digital games is based on the fact that everyday perceptions of morals and norms apply to digital worlds as they do in real life. Violent games, especially, allow the player to perform actions that undermine usual moral standards, at worst arbitrarily killing virtual characters without having to fear sanctions in real life (Smith et al., 2003). Will Slocombe's article explores the 'inhuman ethics' of the video games *Redemption* and *Bloodlines*. In addition, the spectrum of topics ranges from semiotic analysis of the negotiated ethics when a player's subjectivity is represented by the act of playing (Massimo Maietti), to questions of moral management that assume that players of violent computer games regulate their moral cognitions in order to maintain the pleasure of playing (Christoph Klimmt et al.)

Part V: Politics and Ideology

Apart from war and violence, ethics and morals, questions of politics and ideology undeniably rank among the central research fields of critical game studies. Primarily, the complexity and impressions of realism in current game worlds foreground relevant questions of ideologies, designs of society, political ideas and (critical) implications of the politics of representation that are communicated in and by computer games, as well as in their marketing strategies.

In this sense, Sven O. Cavalcanti analyses the dreary nightmarish computer-games worlds in regard to the ways they playfully anticipate the potential failure or complete collapse of modern societies. Andreas Jahn-Sudmann and Ralf Stockmann look at computer games such as *GTA San Andreas* (2004) as cultural articulations of the politically incorrect. And Maia Engeli shows how the modification and re-contextualization of first-person shooters has been established as an art genre within which particularly politically motivated works are developed.

Part VI: Computer Game Play(ers) and Cultural Identities

Cultural identities play a crucial role in actual games studies. The present worlds of the MMORPGs assemble players from all over the world simultaneously and open up a multitude of possibilities to test and negotiate cultural identities at the interface of game world and player's world. Games and game play, accordingly, involve a global element and transcend geographical, cultural and social boundaries, while at the same time they remain situated in local and very specific sociocultural contexts. The production, utilization and status of games in Europe, for instance, are significantly different from Asian game culture's cultural dynamics. A key feature of the East Asian online games context is the relationship between the use of regional aesthetic and narrative forms in game content and the parallel growth in regionally focused marketing and distribution initiatives. Dean Chan examines how this sense of 'Asianness', or intra-Asian cultural identification, is manufactured and marketed in East Asian MMORPGs. In another article, Lynda Dyson explores the way *The Sims* has been used by a group of London girls. Drawing on interview material, it suggests the game provides a private arena for 'acting out' scenarios and fantasies based on social relations in the girls' lives. *The Sims* gives them the means to explore their lives and relationships and potential transgressions through play at a time in their lives when they are experiencing significant levels of surveillance and anxiety from adults. Finally, Dean Lockwood and Tony Richards, in contrast with many theorists who, under the influence of Althusser, point out the 'apparatus' nature of the player-game interface, emphasize the active and non-immersive nature of the game space and advocate an understanding of computer games as an identity-*challenging* space, an identity-'*un*deciding' space, and a 'playing-machine-medium'.

Part I
Game Design and Aesthetics

1
The Aesthetic Vocabulary of Video Games

Joost van Dreunen

'... to what extent can culture itself be characterized as play ...'
Jan Huizinga, *Homo Ludens* (1938)

The social relevance of media technological development lies in its potential to (re)introduce novel elements to human communication and cultural expression. With a greater variety of tools at their disposal, people are able to understand, manipulate and express their experience of reality in different ways. This in turn may have dramatic effects on the underlying principles of social coherence: 'the principal effect of media technology is on social organization' (Carey, 1988, p. 302). Ultimately, how we communicate tells about how we relate.

In recent years, video games have grown into a defining cultural element. In interpreting this new phenomenon, tropes such as Murray's (1997, p. 185) 'participatory narratives', Aarseth's (1997, p. 1) 'ergodic' texts, or Frasca's (2001) focus on sociopolitical game mechanics are now required reading. These theoretical strands all agree on the meaningfulness of the video-game experience, yet do not extensively comment on its *visual* foundation. In this context, it is important to distinguish between *electronic* or *computer* games, referring to electronically enhanced or computer-based games, and *video* games, characterized by a screen through which a player interfaces with the game (world). The central role of 'video' in game play warrants a discussion of the aesthetic characteristics integral to the larger economy of syntactic elements and representative of the principles underlying and guiding its communicative potential. Because video games mediate experience, the building blocks of their visual dimension contain important clues that reflect on the social organization of contemporary society. By identifying some basic visual elements of typical mainstream video-game imagery, we may begin to answer the

3

question: what is it exactly that we see when we see it, and what does this say about us?

A medium, by definition, represents an externality through which, by internalizing its logic or grammar, people can communicate. In strict sociological terms, the analysis of media concerns itself with the relations between people. More precisely, it is concerned with the social conventions and media technological conditions that facilitate social cohesion. Therefore, describing the visual vocabulary of video games is not merely an exposé of contemporary aesthetics, but is an investigation into the underlying patterns of social interaction. Following Aarseth (1997, p. 17): 'The emerging new-media technologies are not important in themselves, nor as alternatives to older media, but should be studied for what they can tell us about the principles and evolution of human communication'. Subsequently, in order to understand how video games fit into the existing spectrum of cultural expression, we must investigate the elements of the vocabulary that need to be internalized in order to participate meaningfully with(in) it.

Contemporary media

Reading requires mastery of the alphabet. Similarly, video games deploy predominantly (moving) imagery, requiring an understanding of its particular visual syntax, grammar and vocabulary. This epistemological demand simultaneously facilitates and influences communicative exchange because of its particular organizational logic, or its 'incarnate grammatical order' (Dewey, 1958, p. 172). Not surprisingly, the prevalent modes of communication and the organizational logic of society are intimately related (Innis, 2003). Traditionally, in order to understand a particular historical moment in social organization and 'style of life', one studies the mediating social and technological conventions, such as capitalism or the introduction of standardized time (Simmel, 1990, p. 429; Gallison, 2003).

A fertile place for investigating the organizational logic underlying video-game syntax lies in the dialectical relation between a customary media technology and the social conventions that constitute a communicative exchange. Contracting the key concepts – medium and dialect – produces a 'medialect'. The function, meaning and significance of a medium changes constantly because of its adherence to the fluid practices of both social convention and technological development. Succinctly speaking, daily life – how it is experienced and expressed – reverberates within any language, whether oral, written or otherwise. Thus, social groups formulate their own grammar (sociolect), as does an individual (idiolect), in accordance with a rule-set analogous to that of a geographically located dialect. And, in this process, technologies facilitate communication and cultural expression in unpredictable ways. The term medialect then refers not merely to the dynamic exchange between technology and communicative expression, but to the very principle that language is irrevocably fluid, irrespective of the attempts to control, standardize and legitimize it (Rossiter, 2003). Finally, the popularization of a new syntax builds upon the existing ones and in the process eclipses them.

A medialect equally deploys both a socially constructed grammar and pervasive use of mediating technology, since both impose a particular combination of affordances and restraints. The term medialect, as a theoretical construct, corresponds to 'technological interactionism' in which 'social processes are not influenced by technological developments (as in technological determinism), nor are they solely controlled by human negotiations (social constructivism), but by both' (Raessens, 2005, p. 379). This bilateral process constitutes the external conditions under which communication takes place, and is pervasive throughout the exchange itself. 'The aesthetic dimension of new media resides in the processes – the ways of doing, the recombination of relations, the figural dismantling of action – that constitute the abstraction of the social' (Rossiter, 2003, p. 105).

In summation, meaningful exchange within a particular medialect demands the internalization of its grammatical rule; subsequent changes or the emergence of an entirely new one may indicate a structural transformation in communicative exchange and social organization.

The three preliminary concepts central to the medialect of video games are separation, spectacle and speed.

'Mon désir est là sur quoi je tire'[1]

Contemporary life is characterized by the plethora of methods through which we are able to extend (fragments of) ourselves beyond our spatial and temporal boundaries. McLuhan (1962) referred to this as the reorganization of the senses, which subsequently alters our experience of reality and has profound social implications. For example, in ostracizing others with our portable music players we open up to and close off certain influences, thereby creating a particular sensory environment and experience. This separation, a self-conscious manipulation or reorganization of sensory input, occurs in video games in two important ways.

Firstly, the visually disconnected position of the player is essential to game-play. More precisely, the separation from the *events* on the screen enables the player to see where she normally cannot: in other words, this is a cultivated way of seeing. Classics such as *Asteroids* (1981) primarily feature a two-dimensional perspective in which a player has a complete overview of the game environment at all times. Following the environmental settings extends into side-scrollers: the background and foreground appear on the right and disappear on the left (*Zaxxon*, 1983), thereby creating an entirely new game mechanic. And finally, the first- and third-person perspectives place the player much closer to the avatar's point of perception, but avoid surrendering the ability to see where the avatar cannot (*Manhunt*, 2003).

While there are many exceptions to this rough developmental synopsis of 'gamic' perspective (borrowing from Galloway, 2006) it serves to underline how the visual separation between a player and the (avatar within the) game is central to game-play. Driving around in *Grand Theft Auto* (*GTA*) you often have to look 'behind you' in order to successfully estimate the next manoeuvre in a race, chase or flight. Despite its much more detailed environment, perspective in *GTA* is essentially no different from perspective in *Asteroids*: the player can see where the avatar cannot, and internalizing this visual logic is crucial to successfully playing the game.

Secondly, separation from *consequence* is fundamental to a gaming experience. A game in which one is unable to play, lose and play again is not a game. This principle is clearly not exclusive to video games: a televised quiz show offers an obvious parallel. In both cases we participate without suffering the consequences of defeat or victory. This separation is central to the experience and resonates strongly with theatrical theory. In Aristotelian tragedy, for example, 'the spectator has the great advantage of having erred only vicariously: he does not really pay for it' (Boal, 1985, p. 14). Similarly, the popularity of quiz shows does not rely on the *contestants'* success or defeat, but on the *audience's* ability to get a question wrong and keep on playing. Many other visual experiences incorporate a similar logic. The 'horror' genre in cinema requires separation to facilitate experiencing the exhilarating trauma of dying, again and again. Mainstream films such as *Groundhog Day (1993)* and *Run Lola, Run* (1998) also incorporate this ongoing 'enactment of [a] denial of death' (Murray, 1997, p. 175).

Much creative and artistic effort is being spent in testing this boundary. One example is the *PainStation* (2001), which delivers electric shocks to a losing player in an attempt to make the game more 'consequential'

(*PainStation* website, n.d.) Although there are examples of games where the death of an avatar is a nuisance (*World of Warcraft*, 2004), rather than without consequence, it lies in the very nature of video games to be able to start over (to some extent) unconditionally.

The argument here is that visual separation and the frivolous nature of playing are epistemologically consistent, placing video games in a peculiar relation to traditional media technologies. To further illustrate this we must take a step back from the screen and look at what we are seeing.

Video ergo sum

Spectacle is a second characteristic of the video-game vocabulary: the mediated phantasmagoria that transcends any informational value and instead becomes primarily a style, a way of speaking. Historically, there exists a strong relationship between mediation and spectacle. The inventions of the stereograph and the development of the railroad system in the early 19th century, for example, were instrumental in the creation of the panorama (Schivelbusch, 1986, p. 52). The coalescence of these two events facilitated an indulgence in scenery far beyond the quotidian limitations of time and space. Not surprisingly, the widespread popularity of the stereograph vis-à-vis the telescope and microscope, which had a more specialized audience, did not depend so much on its *informational* value, but rather on its ability to provide dazzling visual escapades. The Great Fire of Chicago is considered partially responsible for the popularity of the stereoscope (*The Great Chicago Fire and The Web of Memory*, 1996). Thus, spectacle has been a historic driving force behind a cascading elaboration of visual technologies.

In this context, video games are a next-media technological incarnation designed to appeal to our visual appetite, fuelling the demand for increasing technical capacity of consoles and computers. Looking at the mechanics of each rendition of *Final Fantasy* (*FF*), for example, we must conclude that its game system has only marginally changed over the years. Aesthetically, however, the series continues to evolve. No doubt, the most recent version of the *FF*-series on the PS3 will feature even more stunning graphics but remain loyal to its original rule system. It is the diversity of kinetic spectacle, rather than the function of 'state machines' (Juul, 2005, p. 56), that contributes to existing cultural expression: mediated experiences focus on spectacle as much as on outcome.

In addition to the aforementioned shift from 2D to 3D, greater technical capacity may also revamp existing titles and invigorate game play.

Shinobi, first launched in 1987, has had eight incarnations; it shifted towards a third-person perspective in 2003, and has consistently elaborated its special effects. Likewise *Prince of Persia, GTA* and *Duke Nukem* made the switch from 2D to 3D and currently feature vast panoramic landscapes and dazzling effects.

Again, this development is not exclusive to video games. Cinema also has undergone dramatic visual changes. The emergence of the 'disaster' genre, with movies such as *Dante's Peak* (1997) and *The Day After Tomorrow* (2004), shows a trend towards greater spectacle. The large-scale effects of *The Towering Inferno* (1974) are now relatively tame compared to its more up-to-date counterparts. Until the appearance of *Braveheart* (1995), *Independence Day* (1996) and *Troy* (2004), the epic battle scenes in *Spartacus* (1960) remained largely unrivalled. The increasing reliance on spectacle is most obvious in the various episodes in the *Star Wars* saga. Nonetheless, spectacle is not a static composition that stands on itself, merely depending on colour and detail; it is also closely related to *speed*.

Velocity of imagery

If spectacle overshadows the transmission of information as its primary role, so too does *speed*. The tempo of a classic black-and-white movie, remarkably slower than that of contemporary cinema, illustrates the ongoing acceleration in the velocity of imagery. In the visualisation of speed we find the cultural expression of the technological influence on the pace of everyday life (Kern, 1983). Analyzing the changing rural landscape after the introduction of the automobile, Walter Benjamin (1969, p. 250) lamented: 'film is the art form that is in keeping with the increased threat to his life which modern man has to face.' Put differently, we acclimatize to the acceleration of daily life by visualizing it.

In the context of a media-saturated culture, however, velocity refers to both acceleration and deceleration, and to their different narrative functions. Borrowing from sports broadcasts, both cinema and video games have incorporated *slow motion*. In several scenes in the movie *The Matrix* the momentum with which characters move is visualized by depicting them as moving very slowly. This technique of using varying velocities as different narrative modes is a cultivated way-of-seeing. Machine-enhanced vision transcends the aesthetic boundaries of exclusively informational media technologies and is integral to game mechanics and narrative structure.

Many video games deploy this elasticity of speed as part of their gameplay. The controversial game *JFK Reloaded* (2004) simulates the assassination of the US President John F. Kennedy, including a subsequent replay using a 'time-line slider' to watch it at the 'speed of your choice'.[2] Such a 'slow motion' ability gives the player an edge. The manipulation of 'diegetic' time is central to the game mechanics of *Max Payne* (2001), *Dead to Rights* (2002) and *Gungrave: Overdose* (2004). In *Prince of Persia: Sands of Time* (2003) this correspondence between temporal and visual elasticity is incorporated into the storyline in the form of the main character's 'Dagger of Time'.

In this way speed constitutes a third element of the video-game vocabulary that connects to and elaborates on existing communicative practices.

Preliminary implications

If we add up these three syntactical elements, we can start to consider the implications of the video-games vocabulary. The epistemological consistency between visual separation and the frivolity of play, the communicative style of spectacle, and the elasticity of speed all contribute to a unique sensory experience. While none of these elements are necessarily new or exclusive to video games, it is their coalescence that suggests a unique cultural moment. Paraphrasing Crary (1992, p. 1), the medialect of video games implies 'a sweeping reconfiguration of relations between an observing subject and modes of representation that effectively nullifies most of the culturally established meanings of the terms *observer* and *representation*'. Such changes may indicate a structural transformation in contemporary communicative exchange.

With regards to the first part of our question – what is it that we see? – we must reconsider some of the traditional notions we hold with regard to an 'audience'. Instead of a consumer, reader or receiver, a *gamer* is more of a composer than anything else. The inherently playful, detached vocabulary of video games allows neither the gamer nor the game designer complete control over the experience. This establishes a novel exchange between creator and consumer, producer and receiver, sender and reader. More poetically described, playing a game is closer to 'living inside a symphony than reading a book or watching a movie' (Gee, 2005).

Abandoning the sequential order to create a private narrative is integral to media consumption (Barthes, 1975). Replaying a particular media fragment remains within the traditional 'reading practices' (Jenkins, 1992, p. 17) such as those sanctioned by DVDs. The creator or author remains the sole proprietor of the aesthetic experience. However, a game designer merely hands a player the raw materials with which to create or destroy harmony within the game reality. This is not unlike how a soccer coach deploys players or an army commander his soldiers. Thus, 'not only the design and production of a computer game, but also its reception and consumption has to be considered an active, interpretive and social event' (Raessens and Goldstein, 2005, p. 375).

In so-called 'resource management' games, such as *Command & Conquer* (*C&C*, series) or *Civilization* (series), players create a unique visual experience while remaining within the game's narrative structure (for example, to defeat an opponent). The deployment of buildings and mobile units in these games may not deviate from the storyline but a player can create an unanticipated variation of the experience within the game world. '[Y]ou never step into the same video game twice' (Frasca in Perron and Wolf, 2003, p. 227). As part of the game, players build settlements to generate the troops necessary for the successful completion of the game. In this process the aesthetic experience becomes as important as the narrative, because the logic and outcome of a game are indifferent to a player's aesthetic *modus operandi*. The creation and successful management of an army or character eclipses the teleology of its game world, precisely because one can beat the game with *any* combination of aesthetic characteristics. Video games incorporate a creative dimension that represents the 'cultural layer' (Manovich, 2001, p. 46). In computer code such aesthetic differences are meaningless, but they are not on the level of human interface and experience (Johnson, 1997). So *what we see* is a 'cultural interface', because 'we are no longer interfacing to a computer but to culture encoded in digital form' (Manovich, 2001, p. 70). On a micro-level, the extensive character grooming – the continuous organization of the abilities, items, weapons, etc. of an avatar – particularly in role-playing games (RPG) and SIM games, is equal to, if not more important than, meeting the requirements for successful completion of the game. On a macro-level, the uproar following the *GTA* 'Hot Coffee' mod and the uproar surrounding *Super Columbine Massacre RPG* exposes the sociopolitical tensions underlying a simple aesthetic recombination within software architecture.

To answer the second part of our question – what does this say about us? – popularization of the medialect of video games indicates a structural transformation in the ways that people are able to express themselves. The aesthetic vocabulary of video games incorporates detachment, playfulness and fluidity vis-à-vis the hegemonic rigidity of traditional modes of expression. To paraphrase Crary (1992, p. 22) once more, this transition entails 'a radically different visual language that cannot be submitted to the same methods of analysis, that cannot be made to speak in the same ways'.

The examples are many. The emergence of the 'Machinima' phenomenon – the use of software architecture of video games in creating short cinematographic narratives – suggests that 'audiences' have started to manipulate (elements from) their media environments. This is not a purely aesthetic exercise, but through modification and emulation game mechanics facilitate the manifestation of latent sociopolitical realities. The quintessential example is the mod that became a game of its own: *Counter-Strike* (2000). Cheats, glitches and emulation allow radical alterations of the experience that the creators initially may have intended. Fans of *C&C: Generals* (2005) have created a multitude of additional maps, missions and mods that are situated in the context of the current war in Iraq (Command & Conquer DEN, n.d.). Taking this to its logical conclusion, the recent phenomenon of 'counter-gaming', such as *Velvet-Strike*[3], is at its most benign the outcome of a healthy process of explorative curiosity, and at its most political a reclamation of space and the meanings that exist within it. Similarly, 'it takes a game like *Special Force* [a first-person shooter (FPS) based on the armed Islamic movement in South Lebanon], with all of Hezbollah's terror in the background, to see the stark, gruesome reality of *America's Army* [an FPS describing the experience of a soldier in the US Army] in the foreground' (Galloway, 2004).

Video games represent a vocabulary that (re)introduces a degree of playfulness into the process of communicative exchange, thereby facilitating a greater variety of different 'readings' and the manifestation of individual (cultural) expression. Without a doubt its medialect will be eclipsed by a new vocabulary in due time. But today this phenomenon invites us to explore the boundaries and foundational properties of traditional communicative exchange. Who carries the responsibility for gender/race neutrality in a video game? Can we justifiably extend existing copyright into game space? These types of questions speak to the organization of society at large. It is up to all of us to answer them.

Notes

1. 'My desire is aimed for where I fire/target at' (cited by Virilio, 1989, p. 14–15).
2. <http://www.jfkreloaded.com/instructions/>.
3. '*Velvet-Strike* is a collection of [anti-war] spray paints to use as graffiti on the walls, ceiling and floor of the popular network shooter terrorism game *Counter-Strike*.' For more information, see Schleiner (2002).

2
Can Games Get Real? A Closer Look at 'Documentary' Digital Games

Ian Bogost and Cindy Poremba

Digital games are often celebrated for their *realism*, a reference to their visual verisimilitude rather than an association with something actual. As games begin to push past traditional boundaries and contexts, a new genre of sorts has begun to emerge; one that uses real people, places and subjects as its referents. Sometimes called 'documentary games', these works attempt to make some tangible connection to the outside world. In doing so, new issues emerge, not only concerning the representation of real subjects, but on the appropriateness of doing so within a form commonly used for entertainment.

But what is a documentary game? The label seems to be applied loosely to any game that makes reference (however tenuously) to 'the real world' – games from *Enter the Matrix* (2003)[1] to *Medal of Honour: Rising Sun* (2003)[2] to *Civilization* (1991). Is there any real relationship between the documentary form, as we know it from lens-based media, and these games? Or is this a case of simple remediation – an attempt to reconstruct a genre in another media form without sufficient regard to the properties of that medium? Does evaluating games through the lens of documentary, as it were, restrict our understanding, or facilitate new analysis?

If it exists at all, the documentary-game genre is in its infancy. As such, in this work we take a speculative approach, moving from a documentary theory to a model of what games might deserve the name, and finally a proposal for what documentary games might look like in the future. By examining games in terms of documentary conventions, we ask if it is reasonable to call games documentaries in the first place, if such games expand the role of the form from photography and film, and if the title can shed light on the role and reception of the games that adopt the term 'documentary'.

12

Regardless of whether or not we accept the term 'documentary game' as an accurate or useful one, select games have been given this title by journalists, critics, researchers and game developers themselves. Matt Hanson (2004, p. 132) proclaims that in exploring real-world events, digital games are 'finally going beyond the ability to play battles and create historical re-enactments in wargames, and relate more to the areas of subjective documentary and to biopic'. Eddo Stern et al use the term 'documentary video game' (Mirapaul, 2003) in describing their own game *Waco Resurrection* (2003). Tracy Fullerton calls it 'aspirational pre-naming', the wishful adoption of a genre in the hopes that merely naming it will bring it into existence. Clearly, there is a public desire to acknowledge and define select digital games as non-fiction, and a certain prestige in doing so – as if developers were challenging the intellectual status quo in contemporary digital game production.

Whether this is a case of wishful thinking or a legitimate attempt to create a frame for reception is still an open issue. On one hand, 'documentary' is a term with great cultural currency and therefore is useful in establishing context and expectation. Consumers of art and entertainment alike have a developed understanding of the documentary in the 'ordinary' sense of the word: documentaries are non-fictional accounts of the world. They can be found in *Life* and *National Geographic*, not *Glamour* and *People*; on the Discovery and History channels, not ESPN and network broadcast; at IMAX and art theatres, rarely at the local multiplex cinema. However, documentary also implies, almost exclusively, the visual grammar of film and photography, a grammar that may not be appropriate for digital games. Carrying the excess baggage of photography and film into digital games is a risky proposition; it provides an excuse not to ask what expression analogous to documentary film and photography might look like in games. While labelling games as 'documentaries' may establish a frame of reference, it can also obscure the way that games are expressive in ways different from other media. Borrowing the name 'documentary' from these historical precedents is not a substitute for an evolution in the practice of documentary game-making.

More than any other non-fiction mode, film documentary leverages one prominent affordance of lens-based media: its transparency. Transparency closes the perceived distance between the subject and the recorded image. Digital media can adopt this core property of documentary film by incorporating film itself into games, but such a strategy is a rudimentary use of the computational medium. However, digital media have other affordances: notably, the ability to execute processes – to run code, to simulate systems. Documentary film can represent only one instance of the subject: a particular, individual moment in time, human subject or social situation. As audience members watching such films, we are complicit in accepting what we see on screen as the 'definitive moment', as opposed to an aberration. Conversely, digital media offer the possibility of simulating situations difficult to deconstruct or follow visually.

Consider the assassination of John F. Kennedy. In the controversial example of *JFK Reloaded* (2004), players assume the role of Lee Harvey Oswald in an attempt to recreate the series of shots fired in the Kennedy assassination. No matter how we might feel about the provocative nature of such an emotionally charged subject, the game can make a perfectly legitimate claim to the *reality* of its representation – the simulation at the heart of the game and the criteria by which a player's marksmanship is judged are based on real forensic evidence presented to the Warren Commission combined with known physical dynamics such as ballistics. If players have enough skill and luck, arguably they *can* recreate the event as evidenced by this data. We should ask: would the proliferation of conspiracy theories surrounding this event have occurred if the primary cultural document for the event was the accurate physics simulation at the heart of *JFK Reloaded* rather than the grainy and ambiguous image presented in the Zapruder film?[3] Game documentaries offer the opportunity to explore other avenues of non-fiction representation – and in turn, to reveal how our expectations of recording and documentation can be skewed by the myth of cinematic transparency.

Defining documentary games

From the perspective of traditional views of the documentary, the most fundamental question to ask of documentary games is this: can games even be understood as documentary, in the tradition of photography and film? In the documentary pioneer John Grierson's early work (1966, p. 13), documentary is defined as the 'creative treatment of actuality'. Grierson's definition of this then-emerging film genre was quite fluid, and he was inclined to address these works in terms of *documentary quality* rather than assigning an either/or label. We understand two key points at work in this definition. First, documentary quality is not intrinsic to the raw footage in itself. Rather, documentaries are constructed works, and documentary quality is a product of this construction. Second, documentaries are oriented toward the real instead of the imagined, with some intention towards providing insight into the real. As the theorist Bill Nichols (2001, p. 40) notes, 'documentary stimulates epistephilia (a desire to know) in its audience [...] [it] proposes to the audience that the gratification of these desires to know will be their common business'. The completed work is not *primarily* crafted to produce the effect of drama, comedy, horror or any other of the traditional genres of film (although these elements may also appear in documentary). Instead, documentary strives first to satisfy the desire to know *through the authentic representation of lived experience*. If film explores human experience as we expect, wish or fear it might be, documentary cuts through the sheen of this fiction and shows us life as it really plays out.

Nichols' work provides a more contemporary definition of documentary: he notes that documentaries adhere to genre conventions and maintain a certain ethos of production and reception (2001, pp. 22–40). His definition relies on both the producer and the consumer of the work, and suggests the way *documentary quality* becomes a social negotiation between multiple defining factors. For example, material conventions (the use of interviews, a handheld camera, voiceover narration) might be one way for a producer to signal to the audience that what they see on screen is in fact real – although such conventions can be subverted or hijacked to create a 'documentary feel' in fictional works as well.[4] Documentary makers also explicitly index their work as non-fiction through the likes of film promotion, distribution and film festival entries. In this sense, documentary is a flexible form, changing and evolving over time like any genre. A filmmaker's own values, aesthetics and reputation provide additional clues to the work's authenticity and actuality. Deviations from popular standards affect a work's reception as a documentary; so we might see, for example, the works of Michael Moore excluded from 'true' documentary for reasons of character (specifically, an admitted bias and agenda on Moore's part) where current expectation demands an objective, impartial approach. The audience also plays a role in the acceptance of a work as documentary: they approach the work with a 'desire to know' and a primary expectation to be informed as well as entertained. They also assume that the events on screen have actually taken place, in large part at least. Genre conventions and the filmmakers' assurances and reputation help establish this expectation.

From this preliminary exploration into what makes a film documentary, we can define some points of intersection and translation for documentary games. To bear the name 'documentary', games must articulate an actuality. The illusion of transparency facilitates this representation in film, but the constructed space of digital games makes the articulation of actuality more complicated. We must consider that the type of actuality at work in games might be different from that at work in film. In light of the uncertainties of the concept of documentary, we could reasonably reframe the question of 'what counts as a documentary' as an assessment of 'documentary quality'. This approach does better service to the spirit of Grierson's initial definition, simultaneously recognizing games as a different medium and acknowledging that they can make use of real-world referents and remediate the values of lens-based documentary.

Grierson intended to contrast actuality with expectation; documentary is thus a kind of depth analysis that cuts through the hopeful façade of fiction and exposes the world itself. We suggest another way to understand actuality: as the set of everything that does happen or could happen, the overall possibility space for real life. An alternative model might stem from Alexander Galloway's revisionist notion of realism in games, set in contrast to visual verisimilitude, which he calls social realism. Under this understanding, documentary games strive to demonstrate the constraints that produce actual events, to find the limits of human experience and ask what rules constrain that experience such that it takes place in a certain way. For example, a documentary film about a political uprising might focus on the way events really played out in the streets or on the battlefield, or how they were filtered by the Western news media. But a documentary game about the topic might focus on the way local history and politics colluded over time to produce the conflict in the first place. Documentaries excel in specific instances, but documentary games deal in *real* virtualities; possibility spaces in which multiple instantiations for real-world activity can exist.

In his first principles of documentary, Grierson calls for cinema's observational power to bring forth a new art form, one where subjects 'taken from the raw' would give us more of an ability to interpret the modern world. To reframe this call, we must exploit the digital game's potential for exposing the underlying processes of experience for greater understanding of said subjects. Grierson (1966, p. 145) notes: 'In documentary we deal with the actual, and in one sense with the real. But the really real, if I may use that phrase, is something deeper than that. The only reality which counts in the end is the interpretation which is profound.' The illusion of photographic material as *raw* today seems a bit naïve – even within mainstream documentary film. We no longer possess an unwavering faith in the factuality of the photographic image over alternate means of representing reality. And as a flipside to our willingness to question the indexical and iconic claims of film, we are also increasingly willing to extend indexicality to non-filmic media. Fullerton reveals that while computer simulations had previously been restricted in legal proceedings, they are increasingly admitted into evidence. She notes this shift in the evidentiary value of simulations 'is illustrative of how we may someday embrace the possibility of simulations which not only visually model, but behaviorally model aspects of history so that they may constitute "evidence" by that same "social, semiotic process" that gives us the concept of the documentary image' (Fullerton, 2005). It is in the context of this changing perception of the representational qualities of media that the simulations and systems native to digital games can be understood as real – the 'real' process extracted from a given actuality. Digital games can extend beyond lens-based representation in favour of logic-based representation, representing new but equally factual realities by laying bare the logic of a system. In doing so, they can present this deeper reality; Grierson's 'profound interpretation'.

Digital media is hypermediated. Games can contain the same recorded video, photo and audio objects used in film, and in such cases there is little reason to suspect these artifacts would not retain their evidentiary value. The presence of these elements often extends documentary quality to other highly constructed media such as animation, as evidenced by the Academy Award winner *Ryan* (2004)[5], which uses audio interviews with the famed animator-turned-panhandler Ryan Larkin in conjunction with highly subjective 3-D animated visual representations. Games such as *Brothers in Arms: Road to Hill 30* (2005) exhibit higher documentary quality through the use of primary source material, although in the case of *Brothers in Arms*, these traditional documentary elements are ostensibly peripheral to the game-play. In the game, players engage in standard Second World War battle-play, with the potential to unlock additional material (including letters, maps and photos) used as reference in the production of the game – the real material used to create the illusion of realism. *Waco Resurrection*, in which the player uses incantations to control a battle between the Branch Davidian leader David Koresh and the American ATF, also introduces such 'primary source' material; for example, a song produced by Koresh becomes an eerie soundtrack to the game experience. While the inclusion of such media artifacts should not be the sole consideration in identifying a game as documentary, the presence of this material may serve to support the documentary quality of these games. However, like a fictional film that includes stock footage, the mere playback of primary source video and audio artifacts cannot be enough to code a game as a documentary work. Instead of trying to force them to act like film, we must recognize that games document different kinds of actuality.

Forms of the documentary digital game

If digital games maintain sufficient documentary quality to be labelled documentary, but in a different fashion from film, what kind of works could we expect to see in this genre? Again, we can look to Bill Nichols (1991, pp. 12–31) for forms of cinematic documentary, and find several modes defined: the *expository, observational, participatory, reflexive, performative* and *poetic*. Reframing these modes to apply to digital games not only helps define potential growth in the genre, but reveals and reinforces the strengths of each mode. A reinterpreted framework might look as follows:

Procedural: Based on Nichols' expository mode, the procedural mode involves the structuring of the subject in a rhetorical frame produced by a defined rule structure (see Bogost, 2005; Bogost, 2007). The rules become an authoritarian observer, creating the illusion of freedom while defining the scope of the game's constructed actuality. Game examples include *Under Siege* (2005) where players take up the Palestinian cause in Israel, and *Escape from Woomera*, wherein players take the role of a detainee in an infamous Australian refugee camp. In *Under Siege*, the rules define a possibility space that embeds a Palestinian perspective. For example, in one proposed scenario, players would encounter the Israeli Baruch Goldstein, responsible for killing 28 Muslims in a 1994 mosque massacre (Šišler, 2005). The game scenario would draw upon documentation of this factual event to construct its core logic – however, the derived rulesets and possibility space construct a particular perspective on the attack. That the scenario is made interactive in the first place (as opposed to offering the passive experience of reading reports of the incident) suggests intervention is desired, preferable and should take the form prescribed in the game (can Goldstein be deterred? is a violent or nonviolent resolution required?). In *Woomera*, the search for assistance in the camp before deportation plays out as a puzzle game – an often frustrating one that evokes the actual struggle of the camp's refugees. Although in its purest form, the expository mode has waned in cinematic documentary, games might revitalize this genre, particularly if we consider rules as a kind of 'voice-of-God' exposition.

Participatory: Nichols' participatory mode has two aspects: one of embedded observance, and another of situated reception. In traditional documentary, only the documentarian takes part in the participatory mode. But in the game form, players themselves become the participant/observer. To construct this scenario, documentarians may draw from actual experience to draft the rules of a game from it. In the process, they may overcome one of the inherent limitations of film: the impact of the camera. The creators of *Escape from Woomera* built their model of the detention camp from primary research – the game was constructed from images, stories and primary-source documents smuggled from the facility, which was off-limits to visitors and media. In many cases the participatory mode has also accommodated the presence of the construct: acknowledging, in a sense, there is no true observation. This is an advantage for documentary games, as the constructed nature of the experience is never as transparent as in lens-based forms.

Reflexive: The reflexive form of documentary focuses on meta-commentary; it is both a documentary and a critique of the form. This is a mode in which game documentary has strong potential, given players' ability to deconstruct the rules of the system in order to master the game. Games in this genre may also be subject to the overall cultural metaphor of 'game' in relation to reality: simply revealing the 'gameness' of a situation leads automatically to commentary. The game may even work by exploiting the gameness of documentary itself. A meta-documentary such as *Eyewitness* (2003) is reflexive in this sense. In this prototype, the player takes the role of a photographer documenting the atrocities of the massacre of Chinese civilians by Japanese forces in Nanjing. The game lays out a single instantiation of an event, while the player actively takes the position of constructing *a reality* from the historical recreation. Another example of the reflexive form might be found in Imaginary Production's photography game *Snapshot* (forthcoming), wherein players have to figure out how to 'infiltrate' their subject. A more abstract representation such as *Waco Resurrection* also overlaps in this category. The game critiques both the official 'reality' of the ATF siege and the popular 'reality' of games in the current commercial marketplace.

Still using Nichols' work as a contextual frame, we can suggest additional, speculative modes of which few examples presently exist:

Generative: While there is an obvious affinity between lens-based media and raw observation, recording observation is impossible in digital games (there is no equivalent to the camera; all games must build their representation from scratch). One potential reinterpretation of observance might be games driven by real-world data or processes or continuously constructed by its subjects. Such works are prominent in digital art, so it is reasonable to conceive of documentary digital game equivalents. This mode stands in counterpart to Nichols' observational mode.

Poetic: Deeply situated in early modernism, the poetic mode is a more abstract presentation of raw material designed to evoke mood, loose association, and fragmented and subjective perception. A possible work of this type might be Engeli and Czegledy's *Medieval Unreality* (2003), where primary images and sounds are enclosed in a metaphoric game structure as a response to Albania's 'blood feud'.[6] This series of art games stems from a game-modding workshop, where Albanian participants integrated their own documents and histories to produce variations of the game.

Use of the model

In addition to using the categories above for locating existing works identified in the genre, we might take a speculative approach to suggest new opportunities for documentary games. Let's consider sports documentary: certainly there's a difference between a repeat viewing of last night's game, and a work such as *When We Were Kings* (1996). The creative treatment of the work informs its documentary quality, as do the incoming expectations set for the player. For example, the *Madden* series of American football games is not put forth as documentary, nor is it structured as such. But it seems reasonable that one could create a documentary sports game centered on a factual circumstance. A game could present a detailed modelling of actual players' abilities, situated around a real game (this already happens to some extent, but not such that it evokes actuality). In this hypothetical game, the design imperative might be to present a historic series with enough accuracy and specificity to successfully argue that alternate outcomes have a *reality* to them ('If player X would have thrown the ball to player Y, they would have won'). Using live data might produce a *generative* work; whereas allowing players to manipulate the game 'document' (how the sport is presented) might suggest a *reflexive* approach. Or perhaps the *documentary quality* of the game might simply be reinforced through supplementary material – unlocking player interviews, histories, video, news reports and the like.

Implications

When we make a claim for the existence of documentary games – or at least, games with documentary quality – we confront questions of responsibility, rights and media literacy with regard to digital games. Interestingly, these issues are not endemic to documentary games, but may simply be less apparent in traditional works. Does a designer have the right to construct a game from a personal biography that does not necessarily adhere to the historical record (instead arguing multiple instantiations of a claimed actual system of events)? Do we feel as strongly about such a reconstruction as we would a similar cinematic reconstruction? The existence of such games may also underscore deeply rooted views on history, justification and impact. Exposing the underlying systems at play, revealing alternate histories and embedding participants in these experiences is an entirely new model for preserving cultural memory, and not necessarily one willing to uphold the status quo.

Further, we must also examine documentary games in relation to the cultural position of games as a 'low' entertainment medium; games might seem trivial, childish or generally inappropriate as hosts for actuality. We can trace this sentiment to detractors' attacks against those games called 'documentary' in the mainstream media. In reference to the near-universal public condemnation of *JFK Reloaded*, Fullerton (2005) asks why the game suffers so much more grief compared with other media that tackle the same subject: '[...] (W)hat is it about this particular scenario that provokes such strong feelings [...] Do we condemn Oliver Stone's *JFK*? The History Channel's *The Men Who Killed Kennedy*? Or any of the innumerable books, websites, reports, documentaries and other forms of discourse surrounding this event?'

Fullerton puts this view of games as 'low culture' at the heart of attacks on the genre. This view will likely change over time, but defenders of documentary games must also recommend ways in which the medium can personalize and engage participants in real events in new ways. Documentary games reveal new knowledge about the world by exposing underlying systems and embedding participants in those systems. Digital games are a popular and powerful medium with a potential yet to be fully explored, and an area in which actuality and documentary might still find a place. But to enjoy further success, they must move beyond the mere instantiation of 'documentary' as a legacy, and work to define the properties endemic to the genre in digital game form.

Notes

1. Suggested by Hanson (2004, 135).
2. Suggested by Fullerton (2005).
3. A recent Gallup poll shows just 19 per cent of Americans believe the Warren Commission's conclusion that Lee Harvey Oswald acted alone when he killed John F. Kennedy (Gallup Organization, 2003).
4. One commonly cited example is the use of fake 'primary source footage' in Oliver Stone's 1991 film *JFK*.
5. Best Short Film, Animated (2005).
6. Using the original game engine and rules of *Unreal Tournament* (1999).

3
Emotional Design of Computer Games and Fiction Films

Doris C. Rusch

Introduction

The majority of computer games share certain traits with fiction films. Both media often cover common ground in terms of genre, themes, settings, the aesthetics of audiovisual representation and cultural references. Thus, comparing them seems the obvious thing to do in order to learn more about their similarities and differences.[1] But until now, such intermedial analyses have focused mainly on the structural characteristics of games and films (Juul, 2001b), with the result that games appeared to be so different from films that it was easy to conclude that nothing could possibly be learnt from the older medium for the further development of the newer one. This article focuses on the emotional experience of games and films. According to Tan (1996, p. 46), emotion is defined as 'a change in action readiness as a result of the subject's appraisal of the situation or event'. The fiction film evokes emotional experiences by appealing to certain source concerns of the audience, such as security, love and freedom, which are endangered in the course of the narrative. The wish to restore the desirable states that result from the fulfilment of the source concerns promotes action readiness. This is basically true for computer games also. But in addition to the source concerns addressed on the level of narrative, games also appeal to game-specific source concerns such as agency and the feeling of sensorimotor or cognitive competency. Combining these two kinds of source concerns in a computer game potentially elicits a much more complex emotional experience than that elicited by watching a fiction film.

In the following, I will introduce four key pleasures that constitute the emotional experience of both film-viewing and game-playing. These key pleasures shall be referred to as the *visceral eye*, the *vicarious eye*, the *knowing eye* and the *voyeuristic eye*. My hypothesis is that games and films provide their audiences with the same range of pleasures; the differences in the emotional experiences they allow for are a result of the extent to which particular pleasures are provided and of the means with which they are evoked in different media, rather than a matter of principle.

Also, in the following a model of emotional design will be introduced that cross-medially investigates the interrelationships between the aforementioned pleasures on the various levels on which games and films operate. This model has been developed with single-player games in mind. Though it basically still works for multiplayer games, it does not account for the factor of social interaction. The model therefore cannot be expected to exhaustively explain the emotional experiences provided by multiplayer games.

The model of emotional design

This model considers only those medial elements that constitute the emotional experience of the viewers/players. For a more encompassing understanding of how games and films elicit emotions in their users, the media experience and sociopsychological predispositions of these users would have to be taken into account, but this exceeds the range of this article. The model of emotional design:

1. Allows comparison between computer games and fiction films without compromising the uniqueness of either medium or reducing it to only one operational level (like the story or the game play).
2. Clarifies the similarities and differences between computer games and fiction films from an experiential perspective.
3. Helps to identify potential for development in computer games concerning the emotional range and depth of play experiences.
4. Contributes to a better understanding of the nature of computer games and serves as an instrument for the critical analysis of computer games.

Three levels and four pleasures

Starting with the hypothesis that the range and depth of a film's emotional experiences are created by four key pleasures and that these key pleasures can also be provided by computer games (although they are differently emphasized due to medial characteristics), I aim to understand how they work in films and how they can be integrated into games. As Torben Grodal points out in his essay about the pleasures of control, simply transferring a situation that has a certain effect in a film into a game, hoping to achieve the same effect, will not work:

> When viewing a film the labelling of the emotions felt is determined by the viewer's passive appreciation of the film character's coping potentials. But when the situation is part of a video game, it is the player's assessment of his own coping potentials that determines the emotional experience (Grodal, 2000, p. 201).

However, that does not mean that certain kinds of experiences are restricted to a certain medium. They just afford different medial strategies. This is due to the fact that film creates its various pleasures on only two levels – the level of narrative and the level of the interface – whereas the computer game additionally has to account for the essential level of the rule system.[2] Adapting a filmic pleasure strategy for a computer game always has to acknowledge the affordances of the interactive medium. This means that if you want the player to experience something emotionally, staking only on the level of narrative will not suffice. You have to make use of the rule system to support the narrative message you want to convey with the game-play.

The operational levels of games and films

The level of narrative. Being concerned with the experiential rather than structural aspects of games and films, the primary question relates to what function the level of narrative serves in both media. Basically it provides the players/protagonists with goals, conflict and an element of uncertainty. From these narrative cornerstones arises dramatic tension, an experience that is sought and encountered in computer games as well as fiction films. The main difference is that the most dominant basic paradigm of narrative[3] in films is *character*, meaning that the goals and conflicts are shaped by the protagonists' psychological struggles, whereas in games the most dominant basic paradigm is space, so the goals and conflicts are strongly related to the narrative architecture of a computer game (Jenkins, 2004).

Narrative experience in games can take two forms. There is the embedded narrative that consists of the prescribed moments and structures that are relatively fixed in the game. But the true strength of games on the narrative level lies on the emergent narrative, the narrative that arises during play often in unexpected ways, as a result of the individual moment-to-moment game-play (Zimmerman and Salen, 2004, p. 383).

The level of interface. In films, the interface is the audiovisual representation of the narrative. Cinematic techniques shape style and tone and add connotative meaning. In games, the level of interface serves the additional function of communicating between game-system and player. It can convey information acoustically, visually and haptically, providing the player with a feedback to his or her actions and with clues about what to do. As Steven Poole (2000, p. 178) notes, 'the game screen is inscrutable when approached as simple representation; it demands to be read as a symbolic system'. System feedback plays an important role in the emotional engagement of the player; for example, seeing the health bar drop rapidly or feeling the controller rattle will evoke a change in action readiness. The type of interface (cluttered with status bars or devoid of them) has a huge influence on the emotional experience of playing.

The level of rule system. The level of rule system (in the sense of defining interaction) is unique to games. It determines the actual game-play and is in most computer games not directly accessible by the player but can only be deduced by interpreting the feedback the interface provides to the player's actions. The level of rule system is inseparable from the other two levels. It defines what system information has to be displayed on the level of interface, and the way it is intertwined with the level of narrative has a major influence on the emotional experience for players. If you want to get a narrative message across, couple it to the rule system. Here, the ways are defined in which a certain narrative element can be acted out until it is not only understood but felt by the player. Not only does the narrative give extra meaning to the rules (Zimmerman and Salen, 2004, p. 387), but the rules help to perceive the narrative. There will be examples of this later on.

Categories of pleasure in games and films

Jon Boorstin, whose observations about what makes movies work serve as the main source of inspiration here, uses the metaphor of the *visceral eye*, the *vicarious eye* and the *voyeur's eye* with which a recipient watches a movie, to categorize the pleasures a film provides to the viewer (Boorstin, 1995). In order to include the important joy of *thinking*, I added the category of the *knowing eye*. (For the sake of brevity, the 'eyes' will stand as representatives of the pleasures they provide for the viewers/players.)

As the emotional experience of watching a film or playing a game is a team effort of the various eyes, it is often hard to say which eye causes a particular emotional response from the viewer/player. But although the eyes need one another to create an emotionally satisfying experience, one or two often dominate the others, depending on medium (game or film), genre (romance, horror) and particular medial realization (style and tone). This means that the categorizations cannot be as clear-cut as some might feel they should be. Also, it is not my intention to explain all aspects of the four eyes exhaustively, but rather to convey the idea about how they work on the various operational levels of games and films.

The knowing eye. This first pleasure category aims at getting the viewer/ player involved in the film or game through intellectual stimulation. On the level of narrative, its task is to follow the plot in order to build hypotheses about what is going to happen next. One of the pleasures it provides is the pleasure of being right or – because a totally predictable plot easily becomes dull – being surprised with a clever twist in the narrative. The knowing eye is proud and does not like to be duped. It insists on plausibility and is incorruptible by overwhelming images or bombastic sounds. Even if some of the other eyes are completely overawed, if the knowing eye is ignored, it will spoil the fun and whisper, 'Nice, yes, but nonsense'.

The knowing eye also connects the content of a film or game to rest-of-life knowledge, contextualizing the media experience and drawing upon knowledge about a certain genre and its specific rules, using this knowledge to build more accurate hypotheses about the course of events. Postmodern films often challenge this knowledge and the pleasures they generate are mostly pleasures of the knowing eye, because before you can appreciate creative rule-bending you have to know the rules.

On the level of interface, the knowing eye derives pleasure from interpreting connotations established by the cinematic codes. If it becomes too dominant, it blinds the other eyes, making watching a movie a purely intellectual experience. That is not necessarily bad, but not what film – at least the classical Hollywood fiction film – is all about.

In current games, there is not much for the knowing eye on the narrative level. Even if a game has a back-story, it rarely provides the player with enough narrative information to build hypotheses about what is going to happen next, story-wise. One exception is the innovative game *Fahrenheit* (2005) that works more like an interactive movie than a computer game. Another one is the Sony PlayStation game *Ico* (2001),[4] because here the narrative is tied to the dominant basic paradigm of games: space. The game is about the escape of Ico and Yorda from a gigantic castle. That the player can only travel in one direction is a prerequisite for the dramatic design of the narrative architecture. Early in the game, Ico and Yorda arrive at the main gates of the castle only to find them closing when they advance. This establishes a long-term goal of the game. As the player progresses, the castle's architecture becomes more and more transparent, allowing for hypothesis-building about the course of the journey. Seeing the drawbridges that cannot be reached at the moment but will have to be let down later, and being able to speculate where they might lead, has a dramatic pull effect that gets stronger the closer one gets to the end.

The reason why following the narrative of a game is often tedious is that in most games it is completely separated from the game-play and conveyed mostly through cutscenes. A solution to this problem would be to relate more strongly the narrative level to the level of the rule system, thus focusing more on emergent than embedded narrative (see above). This is done in *Ico*, where the narrative is inseparable from the player's actions. Having to protect and rescue Princess Yorda is not just an abstract goal of the narrative, it is what the player does. For example, you cannot leave Yorda alone for too long or the shadow demons will get her, then the world freezes and the game is over. The rule system turns the romantic metaphor of two people being unable to live without each other into cold game-play reality.

But *Ico* is an exception in many respects. In most games, the knowing eye is more focused on the level of interface, where it deciphers the game world and interprets its signs as clues for the player on how to act. The level of interface is where the knowing eye accesses the level of rule system. Thus, the hermeneutic process that takes place on the level of interface in games is different from that in films. Steven Poole (2000, p. 185) subdivides it into two parts: 'Imagining into' and 'imagining how'. 'Imagining how' because at every moment this operation precedes the dynamic challenge of being able to predict how one's actions will affect the system, and therefore what course of action is optimal; 'imagining into' because one needs to understand the rules of the semiotic system presented, and act as if those rules, and not the rules of the real world, applied to oneself. The requirement is to project the active (rather than just the spectating) consciousness into the semiotic realm. The video-game player is absorbed by the system: for the duration of the game, he lives among signs (another way of describing the dissolution of self-consciousness in the video-game experience).

Deciphering the rules of highly complex games such as *EverQuest 2* (2004) can be one of the key pleasures of playing. A highlight for the knowing eye on the level of rule system is to find the gaps in the system that allow the player to solve problems in a way not anticipated by the designers.

The vicarious eye. Following Boorstin (1995, p. 66), 'the vicarious eye puts our heart in the actor's body: we feel what the actor feels, but we judge it for ourselves. The voyeuristic experience may be grand or clever, but the vicarious experience can be profoundly moving'. The key word here is empathy. To feel it, you have to care for the protagonists, be sympathetic to their goals and conflicts and establish a positive disposition towards them. All this is achieved on the level of narrative. Personal information about the characters allows the audience cognitively and emotionally to approach the character's position, in order to understand his or her feelings. Seeing characters you care for struggle in conflict leads to suspense, and suspense is experienced as entertaining (Vorderer and Knobloch, 2000, p. 62). To create suspense, the vicarious eye has to cooperate with the knowing eye, because how can you fear for a character if you are not aware of any danger?

Because most games do not give information that makes the characters psychologically more interesting, the vicarious eye is somewhat neglected in games. Non-player characters (NPCs) are mostly reduced to their functional roles in the game. Therefore, you do not really care about their fates. Again, an exception can be found in *Ico*, where a strong bond between Yorda and the player is established during the game, until rescuing her becomes much more than just the game's objective, but a real motivation to keep playing. You not only want to know how it ends, you want to save her and when you think you have failed, it hurts. The emotional bond that is established between the player and Yorda during the game is – again – due to the pervasive coupling of narrative and game-play. The affectionate gesture of holding Yorda's hand makes you stronger when you have to fight demons. Also, having to pull her out of the portals into which she is dragged by enemies has a much stronger emotional impact than would simply calling for her. On the level of interface, controller behaviour further enhances the vicarious experience. Holding Yorda's hand makes the controller vibrate, suggesting that the player can feel her pulse.

The key to creating a vicarious experience in games is making the player care about the characters. To achieve this goal, one must not concentrate on the level of narrative alone, but find a way to make character traits and relationships between characters meaningful for the game-play,[5] and to use emotionally engaging metaphors to display the relationships between characters on the level of interface.

The visceral eye. In the realm of the visceral, film and game come closest to each other. Experiencing something first-hand is the pleasure the visceral eye provides and this does not need any characters.

When we watch a rollercoaster sequence in a movie, how much of the thrill comes from seeing the riders appearing scared and sharing their emotions, and how much comes from experiencing the ride ourselves? Which would be more effective, shots only of the screaming riders or only point-of-view shots of the rails rushing up at us as we dive and twist and turn? The riders aren't characters in the empathic sense, they are cheerleaders for *our* rollercoaster ride, there to intensify our thrill of motion, but more likely than not they just get in the way (Boorstin, 1995, p. 110).

The visceral eye responds to everything in a film that stirs the beast in the watcher: the appalling face of the mummy, the sensations provided by soft-core porn. In these moments we want the character to be us, we want to experience the moment directly (Boorstin, 1995, p. 111). So, by catering to the visceral eye, the fiction film gives a taste of what games are best at; letting you *be* the character and providing you with first-hand experiences. In games, the visceral joy of first-hand experiences is strongly related to the experience of agency. Even though the visceral and the knowing eye are incompatible, agency needs them both. To experience agency, it must be possible to build reliable hypotheses about action and reaction, but experiencing agency also means viscerally enjoying the outcome of a certain action, such as watching the monster you have just hit with a grenade explode into bits.

The most compelling aspect of the fighting game is the tight visceral match between the game controller and the screen action. A palpable click on the mouse or joystick results in an explosion. It requires very little imaginative effort to enter such a world because the sense of agency is so direct (Murray, 1997, p. 146).

The visceral pleasures happen on the level of interface. Even if in a game the rule system is needed to provide the player with something to experience first-hand in the first place, the fun part is watching (and hearing) the outcomes of the action, to feel the controller rattle in the heat of battle.

The voyeur's eye. This is the eye that looks at the world presented in a film or game from a safe distance. It is the visceral eye's cool brother. Like the knowing eye, it is not easily seduced. 'The voyeur's eye is the mind's eye, not the heart's, the dispassionate observer, watching out of a kind of generic human curiosity. It is not only sceptical, it is easily bored' (Boorstin, 1995, p. 13).

What the voyeur's eye needs can be found on the level of interface. There it takes its pleasure from a richly imagined special world, full of enticing things (Boorstin, 1995, p. 12). In films, this world is brought alive by careful mise-en-scène. In games, the world also needs a credible physique. Wading through water must feel different from walking on solid ground. Games are specialized at providing voyeuristic pleasures due to the fact that their most important narrative paradigm is not character but space.

Conclusion

This article introduced a model of emotional design that identified four key pleasures provided by games and films and investigated how they relate to each other and to the various operational levels of the two media. In reference to Jon Boorstin's work I called those pleasures the *vicarious*, the *visceral* and the *voyeur's eye*, adding the *knowing eye* to include the joy of intellectual stimulation. Whereas films have two operational levels (the narrative and the interface), games possess the additional level of the rule system.

The model shows how the various eyes are differently emphasized in films and current games, revealing potentials for future development. The knowing eye in games is found to be rather 'short-sighted' on the narrative level, leaving potential for the regulation of player interest largely untapped. Instead, games focus on the pleasures of the visceral and voyeuristic eye, but neglect the vicarious eye that is most dominant in film. This corresponds to the different weighting of basic narrative paradigms in games and films that are due to the different modes of reception. Although the satisfaction of game-specific source concerns such as agency and sensorimotor competency (addressed by the visceral and the voyeuristic eye) will always be essential to the pleasure of playing, the emotional range and depth of games can be enhanced by looking more closely at the vicarious eye. The most effective way to do this is to couple narrative elements to the actual game-play. Potential for development in games can further be identified in regard to the knowing eye, because postmodernistic attempts are still very rare but could allow for completely new game concepts. For example, one could imagine a game that works like a Kafka novel, where the rules have to be constantly reinterpreted and the goals fade further away the closer one comes to achieving them. The motivation to keep playing would be the wish to finally grasp the logic of the game world and to beat the game. A broader range of narrative themes, a more elaborated symbolic interface language and a generally stronger interplay of the various operational levels of games are also desirable. Diversity rules!

Notes

1. For an interesting analysis of the relationships between computer games and fiction films, see Poole, 2000, pp. 65–89.
2. This relates to the term 'ergodic', which derives from the Greek words 'ergon' and 'hodos', meaning 'work' and 'path'. Espen Aarseth uses the term 'ergodic', to refer

to any text in which the activity of a 'reader' (partly) determines which signs appear on the surface of the medium. The text that is read from is produced by the recipient's 'work': 'In ergodic literature, nontrivial effort is required to allow the reader to traverse the text. If ergodic literature is to make sense as a concept, there must also be nonergodic literature, where the effort to traverse the text is trivial, with no extranoematic responsibilities placed on the reader except (for example) eye movement and the periodic or arbitrary turning of pages' (Aarseth, 1997, p. 1–2).

3. For more information about the basic paradigms of narrative as *character*, *space*, *time* and *action*, see Mundt (1994, p. 52).

4. A review of *Ico* including a summary of the game's narrative can be found at http://www.gamespot.com/ps2/adventure/ico/review.html.

5. David Freeman (2004) has developed some useful techniques for player-character and NPC bonding.

4

'Applied Game Theory': Innovation, Diversity, Experimentation in Contemporary Game Design

Henry Jenkins and Kurt Squire

Since April 2003, we have written a monthly column, 'Applied Game Theory', for *Computer Games* magazine. Our goal was to try to make some of the core insights of games scholarship more widely accessible to the people who design and play games. Perhaps the hardest challenge of producing this column has not been the issue of how to balance abstract speculation with concrete criticism or how to identify burning topics of interest to gamers; the biggest challenge we faced, as academics, was how to reduce our arguments to 800 words per month without oversimplifying.

During that time, the column has provided us a platform to address some of the core issues impacting the games industries (debates about censorship and media violence, intellectual property, user-generated content, online communities, the serious games movement, the educational potentials of games, and racism, sexism and homophobia). Columns have focused on top-selling commercial games (for example *Animal Crossing*, 2002; *Civilization III*, 2001; *The Sims*, 2000; *Half-Life*, 1998; *Grand Theft Auto 3*, 2001), on independent and experimental games (such as the work produced by Game Lab), and on research prototypes being developed by major universities and media centres around the world.

The following chapter represents a selection of columns focused around issues of experimentation, innovation and diversity in game design. (Many of our efforts on games and education can be accessed at http://www.educationarcade.org.) From our very first efforts, we felt that game critics have an important role to play in educating consumers about cutting-edge work within the medium and challenging the industry to take seriously the potentials of games as a form of artistic expression. Our approach mixed comparisons between games and other media (including in the selection here, film, television, comics and popular music) with efforts to identify properties distinctive to this emerging artform (including a focus on modes of interactivity and participation and on spatiality). We have respected the challenges facing designers working within the games industry (with its relentless focus on the bottom line) but we have also sought to identify ways emerging products (commercial and otherwise) helped to expand the vocabulary of game design.

'Sensory Overload' (July 2003)

This year, the Interactive Digital Software Association hosts its annual Electronic Entertainment Exposition (E3) in Los Angeles. Perhaps you have already started to read advanced publicity for such hot new games as *Star Wars Galaxies, Pitfall Harry, Silent Hill 3, Aliens vs. Predator, Wallace and Gromit*, and *Republic: The Revolution* which will debut in the E3 showroom.

Perhaps you are at the convention now, reading this column over the thundering noise and flashing lights which turn that same showroom into something akin to the streets of Hong Kong at midnight. Scantily-clad floor babes beckon to you with promises of easy access and cheap loot. Dancers in leotards demonstrate the wonders of motion-capture technology. Highly skilled game girls are challenging all comers. The noise you are hearing is the sound of a thousand computer games all being played at the same time. Most people stagger out after only a few minutes, so overwhelmed that they can no longer focus on any one screen. We have seen people passed out in the corner, their friends trying to coax them back to consciousness by upping their caffeine intake. Everyone should see E3 once to experience the adrenaline rush.

E3's economic function is well understood by anyone who has spent more than a few minutes thinking about the games industry. This is where buyers from Wal-Mart Electronic Boutique and other chain stores first encounter the coming year's product. The major game companies are hyping their hottest new titles, smaller companies are trying to break into the market. Both are involved in a life-and-death struggle for the attention of the middlemen who will determine how much shelf space a title will get and how long it remains there. At E3 2001, for example, the disappointing XBox showing sent the Microsoft PR machine scrambling for months to convince retailers that the platform was ready to ship.

Yet, the effects of E3 on the look and feel of contemporary games have been less often discussed. For starters, many game designers talk about the importance of designing memorable moments into their new releases, features which leave vivid impressions after the bulk of what we saw on the floor has blurred together in our sleep-deprived, alcohol-addled and sensorily overloaded minds. Producers push designers to come up with a preview reel which grabs attention on the huge monitors which dot the display room and often the result is an overemphasis on cinematics over game play. The disparity between those massive screens, which would not seem out of place at your average multiplex, and the much smaller monitors on which most of us play games tells us why so many games take on the look of bad action movies rather than exploring the interactive potentials of this medium or why game soundtracks so often emphasize noisy explosions rather than emotionally enhancing music. What would happen if every movie to be released next year all got shown at the same time in the same auditorium? Which films would stand out? Which films would get buried? For those of us who want to promote greater innovation and diversity in game design, the E3 floor may be the biggest obstacle in our path.

Smaller-scale games get little or no floor space. *The Sims*, for example, got swallowed up by the chaos of the E3 showroom. Games such as *Rez* or *Majestic*, which really stretch the limits of our understanding of what the medium can do, are often displayed in private rooms off the main floor. Some of the most interesting games are literally relegated to the basement, the Kentia Hall, where foreign and independent game developers fight over the cheap space with discount distributors and peripheral manufacturers. You might find an interesting title squeezed between the new video-game glove and an online Korean dating game, but these quirky titles have little chance at being heard above the marketing din upstairs.

After even a few minutes on the floor, all of the games start to look the same. Is it any wonder that distributors and retailers are drawn towards recognizable franchises in such a hyperbolic environment? Is it any surprise that retailers make decisions based on eye candy and glitz?

There is nothing wrong with the industry throwing itself a party at an E3. Would not it be great, though, if like film and music, we had other outlets as well: independent gatherings, grassroots festivals, a real awards show. As the games industry matures, it may not be able to contain all of its economic and social functions within one or two gatherings. The Indie Games Jam at the Game Developer's Conference is one approach; we hope that other similar efforts will emerge in the coming years as well. Consider, by comparison, how important the Sundance Film Festival has been for creating visibility and providing economic opportunities for independent filmmakers.

'Refreshing' (October 2003)

Being a game reviewer seems like a dream job: advance copies of games months before they ship and, most importantly, all the free games that you have time to play. Listen to most game critics, though, and you hear that reviewing games for a living can almost take the joy out of gaming. There are only so many dungeons that one can clear or look-alike real-time strategy games you can play before they all, well, start to look alike. Your senses literally become deadened by the repetition of game characters, themes and mechanics.

Even the good games, which can take more than 40 hours to finish, will often throw level after level of monsters at the player with little novelty. How many times do you get a few hours into a game and already know that you have seen it all before and that finishing is more a matter of endurance than excitement?

Fortunately, there are a few gems that suggest ways out of these gaming doldrums. In *Half-Life*, memorable moments are carefully doled out throughout. At first the game surprises the player with its interactive environment; it is not until much later that the player experiences some of the game's other most remarkable features, such as NPC guards that protect the player, marines which at the time redefined state-of-the-art artificial intelligence, or the dramatic desert and surrealist landscapes which appear after hours of being locked in the dark cave-like spaces of the Black Mesa Compound.

Eyecatching graphics or unnervingly good artificial intelligence are sure-fire ways to surprise the player, but games such as *Pokémon Ruby/ Sapphire* for the GameBoy Advance show that good design can also create novelty and surprise. New Pokémon with colourful skills are peppered throughout the game and players delight in 'collecting them all'. Pokémon randomly evolve or gain new skills. And, like *Half-Life*, *Pokémon* introduces new game-play elements such as contests or Pokémon breeding hours into the game, creating the feeling that the game could turn in a new direction at any point.

Even a simple game like *Pokémon Sapphire* reminds us how games can break our expectations, teaching us new ways to think about games as a medium or about the worlds they represent. Media-studies scholars call this process defamiliarization. Our normal perceptions get deadened, much like the poor critic who has to play through the same formulaic games again and again. Art reawakens, refreshes and revitalizes them and encourages us to rethink our assumptions. This is as true for popular art – such as computer and video games – as it is for the so-called fine arts.

A game such as *The Sims* can invite us to rethink our relations with family members or roommates, while a game such as *Half-Life* breaks our expectations about how the first-person-shooter genre operates. Knowing what expectations players have is part of the craft of game design; creatively challenging those expectations without frustrating the player is part of the art.

In both cases, part of what makes these games interesting is how they transport players into entirely new worlds. All media are interactive in one sense – we interpret information from our senses, relate what we are experiencing to what we already know and then build expectations about what will come next. Games are unique in that we act on our assumptions about how the world will operate, putting them to the test. The best designers shatter those expectations without leaving us feeling cheated or lost.

Genres in games, as in other arts, are enabling mechanisms which enhance the communication between artist and consumer, helping us to know what to expect and what we need to do to maximize our pleasure from the experience. The best artist knows when to break with those genres so that they offer us something novel and engaging. In a mature art, we come to read the breaks against the continuities to develop new understandings of the basic thematic building blocks of the medium. The risk is that genres become straightjackets which stifle innovation among artists and deaden the perceptions of consumers. Many game designers protest that the rigid application of genre formulas in the production process, in deciding what games to greenlight, in shaping their marketing, in determining how they get reviewed, and in producing a fairly conservative audience response, is what crushes innovation within the medium. These genre rules are often enforced as powerfully by consumers who are outraged if a first-person shooter does not include such and such a feature. If the medium is to grow, however, both designers and players need to learn when and how they can defamiliarize those formulas to create fresh experiences and to keep us on our toes throughout the duration of game-play.

'A Game That Will Make You Cry' (February 2006)

Want to design a game to make us cry? Study melodrama.

Do not snicker, o ye hardcore gamers. Although we associate melodrama with soap opera – that is, 'girly stuff' – melodrama has historically appealed as much to men as to women. Sports films such as *The Natural* or *Seabiscuit* are classic examples of this, and in fact, most action-oriented genres are rooted in traditions from 19th-century melodrama.

The best contemporary directors of melodrama might include James Cameron, Peter Jackson, Steven Spielberg and John Woo, directors who combine action elements with character moments to generate a constantly high level of emotional engagement. Consider a passage from Cameron's *The Abyss* during which the male and female protagonist find themselves trapped in a rapidly flooding compartment with only one helmet and oxygen tank. Games include puzzles like this all the time, but few have achieved the emotional impact of this sequence.

Cameron deepens the emotional impact of this basic situation through a series of melodramatic devices: playing with gender roles (the woman allows herself to go into hyperthermia in the hope that her ex-husband, the stronger swimmer, can pull her to safety and revive her), dramatic gestures (the look of panic in her face as she starts to drown and the slow plummet of her hand as she gasps her last breath), emotionally amplifying secondary characters (the crew back on the ship who are upset about the woman's choice and work hard to revive her), abrupt shifts of fortune (a last-minute recovery just as we are convinced she is well and truly dead), performance cues (the rasping of the husband's throat as he screams for help), and an overarching emotional logic (she is brought back to life not by scientific equipment, but by human passion as her ex-husband slaps her, demanding that she not accept death). When the scene ends, absorbed audiences gasp because they forgot to breathe. Classic melodrama depends upon 'dynamism', always sustaining the action at the moment of maximum emotional impact.

Critics might argue that these conventions are unique to film, but most melodramatic techniques are within reach of today's game designer. The intensity and scriptedness of a scene like this could not be sustained for 40 hours, but it could be a key sequence driving other events. Classic melodrama understood the need to alternate between downtime and emotional crisis points, using abrupt shifts between emotional tones and tempos to further agitate the spectator. And, we often associate melodrama with impassioned and frenzied speech, yet it could also work purely in pantomime, relying on dramatic gestures and atmospheric design, a technique platform games do well for fun or whimsy (think *Psychonauts*), but few games use for melodramatic effect.

Some of the most emotionally compelling games are beginning to embrace the melodramatic. Take, for example, the now-classic game, *Ico*. The opening sequences work to build sympathy towards the central protagonists and use other elements of the mise-en-scène to amplify what they are feeling at any given moment. The designers exploit the contrasting scales of the characters' small physical builds with the vast expanses of the castles they travel through. The game also relies on highly iconic gestures to communicate the protagonists' vulnerability and concern for each other's wellbeing.

One lesson that game designers could take from classic melodrama is to recognize the vital roles that third-party characters play in reflecting and amplifying the underlying emotions of a sequence. Imagine a scene from a television drama in which a mother and father fight in front of their child. Some of the emotions will be carried by the active characters as they hurl at each other words that express tension and antagonism. But much more is carried by the response of the child, cowering in the corner with fear as the fight intensifies, perhaps giving a hopeful look for reconciliation. Classic melodrama contrasted the actions of the protagonists and antagonists with their impact on more passive characters, helping us to feel a greater stake in what was occurring. Games, historically, have remained so focused on the core conflict that they spend little time developing these kinds of reactive third-party characters, with most NPC seemingly oblivious to what's happening around them.

Finally, the term melodrama originally referred to drama with music, and we often associate melodrama with swelling orchestration. Yet, melodrama also depends on the quality of performer's voices (especially the inarticulate squeaks, grunts and rasps which show the human body pushed beyond endurance) and by other expressive aspects of the soundscape (howling wind, clanking shutters) – elements that survival horror games use to convey fear, but are rarely used for other emotions. Game designers can not expect to achieve melodramatic impact if they continue to shortchange the audio track.

Want to design a game that will make players cry? Study melodrama.

Polyrhythm (January 2004)

If *Arcadia* did not exist, a game theorist might have to invent it. Come to think of it – one did!

When *Arcadia* premiered at the GDC's Indie Games Fest several years ago, it provoked excited response from media scholars, retro-gamers and minimalist game designers. Produced by GameLab, the independent games group headed by the designer/theorist Eric Zimmerman, the co-author (with Katic Saten) of the recently released MIT Press book, *Rules of Play* (Zimmerman and Saten (2004)), *Arcadia* allows players to tackle four games at once – an exploration of the aesthetic and ludic consequences of multitasking. If this were not enough, the games are based on 'classic' Atari-style games, a paean to what many people see as a golden age in game design, an area of experimentation, aesthetic distinction and eloquent game-play.

You can see that the folks at GameLab want us to use *Arcadia* to examine how far we have come as a game-playing culture over the past few decades and to think about some of the core elements of game design. But, push that intellectual pretension aside and *Arcadia* is a darned good little game. If all experimental art were this fun, you'd see people lining up outside the Guggenheim with their pockets full of tokens.

Arcadia confronts players with a random selection of eight games: *Over Drive* (think *Pole Position*), *Tut-Bricks* (*Tetris*), *Scrollius* (*Defender*), *Jumpy-McJump* (*Pitfall*), *Fullclip* (target shooting), *ElectronicTennis* (*Pong*), *"Rocky" Shapiro's Video Baseball* (*Realsports Baseball*), and *Strathreego* (*Connect Four*). Each game is simplified to be playable in only one-fourth of the screen and through a single mouse movement and click. But each still feels like its ancestor. The baseball and jumping games are about timing. The *Tetris* and *Connect Four* games are about patterns and strategy. *Pong* still feels like, well, *Pong*.

The design is crystal clear and yet still evocative. The artists/designers promise to take you back to 1977 and they do, exploiting retro-chic for all it is worth. The boldly colored splash screen could have been taken straight from a 1970s luxury van. The pixilated graphics capture the lovable blockiness of Atari-era characters. The cap-gun sounds of the first-person shooter or the canned crowd cheers after a home run in baseball take the player back to the days of 4-bit gaming.

GameLab creates an entirely new play experience by mixing and matching these familiar materials. *Arcadia* is one part action game. The pace of each game constantly quickens until things fly at you from all directions so quickly that you lose control. *Arcadia* captures what Mihaly Csikszentmihalyi calls 'flow', the idea that we are in a special state of consciousness when all of our senses are engaged in a problem. If you think about what you are doing, you quickly fail, but if you can get into the zone, you can do better than you might otherwise imagine.

Arcadia is one part resource-management game. The game ends when all lives are lost in any one game, so a challenge becomes, 'How do I divide my attention across all four screens to keep them going at once?' Because the final score is the result of multiplying all four scores, different strategies emerge for hanging on a few seconds longer, and scoring evenly in all four games can lead to big payoffs.

The most surprising and instructive aspect of *Arcadia* may be its play with rhythm. *Arcadia* strips each game down to its essential elements – whether it be timing, dodging, pattern-matching or aiming. The player starts by settling into the groove of each individual game. As each new game is added, layer upon layer of rhythm is added to the experience. Here the game becomes truly polyrhythmic. In music, polyrhythm is different rhythms played simultaneously. Unlike, say, a standard rock drumbeat, which might use a snare, kick drum and high-hat to create a relatively unified beat, in polyrhythm underlying rhythms are relatively prime to each other; if picked apart, their basic beats and patterns do not match up at all. Polyrhythm, common in many tribal drumming forms, takes distinct rhythms that cannot be subdivided into matching beats, and layers them on top of one another, creating unique sounds.

What makes *Arcadia* so interesting is the way that it blends the feel of different genre games into one coherent experience. Lots of recent games have played around with rhythm explicitly – from *Dance Dance Revolution* to *Rez* and *Frequency*. In most of those cases, the rhythm is matched to the audio track. Listen to the audio track of someone playing *Arcadia* and you hear something quite different – multi-rhythmic patterns of play from four games, sometimes discordant, other times creating sublime unity.

Arcadia shows how an experimental games industry could exist within the games industry. Released as a Shockwave game, *Arcadia* takes the simple idea of playing four games at once and teaches us some much more powerful lessons about the role of rhythm in game design. We would not be surprised to see these design tricks flow into more traditional commercial games, such as console platform games. To avoid stagnating, the games industry needs to find a way to sustain experiments that push the boundaries of the medium.

'Realism (Does Not Equal) Reality' (December 2004)

Arguments about video games and violence almost inevitably hit on the question of whether, as video-game graphics become ever more realistic, we will reach a point where games are indistinguishable from reality. This is basically the old undergraduate trap of confusing realism and reality.

Realism refers to a goal in the arts to capture some significant aspect of our everyday experiences. No artwork achieves absolute fidelity to the real, and it is pretty extreme to imagine anyone anywhere at anytime confusing art with reality. Realism in the arts, in fact, gets judged as much in terms of its break with existing artistic conventions as it does in terms of how it captures the real. Realism is a moving target, not simply because technologies change, but also because techniques shift.

As a result, nothing dates faster than yesterday's realism. For example, the Italian neorealist films (*Rome: Open City, Bicycle Thieves*) were acclaimed in their own era for their use of non-actors, improvised dialogue, location shooting and episodic structures, all of which were read as creating an unprecedented relationship between cinema and reality; today, viewers groan over their swelling music tracks and reliance on melodramatic clichés. The method acting associated with Marlon Brando in the 1950s was celebrated for its realistic depiction of normal speech, yet again today such performances can seem extraordinarily mannered.

What does this suggest about realism in games? In part, it tells us just where artists are pushing contemporary conventions. Innovations in artificial intelligence might create more natural-seeming non-player characters; 'immersive' interfaces try to situate the interface within the fiction of the world; expansive worlds (such as *Grand Theft Auto*) sell us on the feeling of a setting; accuracy in detail in *Medal of Honor* creates a more realistic depiction of war; realistic physics cause the world to behave in a consistent manner, and photorealistic graphics allow for less cartoonish games.

Almost never does a game design team focus on all of these elements of realism at the same time. They make choices about where realism will achieve the desired aesthetic effect and what needs to be stylized in order to ensure the intensity and immersiveness of the play experience.

In fact, history tells us that most people do not want absolute realism. The Italian neorealist Caesar Zavattini once proposed making a movie which showed 24 hours in the life of characters who did absolutely nothing. If Zavattini were to make a game around that pitch, nobody would buy it. We want games to break with everyday experience. Otherwise, what's the point? Games that embrace a realism aesthetic aspire to create the feeling of playing a role.

People also fear that role-playing in realistic worlds will somehow poison them. But, we have been 'role-playing' in the real world for a long time without obvious detriment. If this were really true, the most dangerous person on the street would be a Shakespearean actor.

In many cases, the realist style may represent a move away from absolute fidelity to the real world: for example, many people read black-and-white and grainy images in film as more realistic than crystal-clear colour images, even though most of us experience the world in colour. Photorealism depends on the representation of camera flare lines which are a property of camera optics rather than reality.

Because we read realism against existing artistic conventions, breakthroughs in realism call attention to themselves – they are literally spectacular accomplishments. When the marines behaved 'realistically' in *Half-Life*, it was so compelling precisely because it was in a game. As long as the artistic devices are foregrounded, we are unlikely to forget that we are playing a game. Realism is not about creating confusion in the mind of the consumer; it is about using the medium to call attention to some aspect of the world around us. And more often that not, the best way to help us see the world from a fresh perspective is through exaggeration or stylization.

Game reformers are not the only people who confuse realism for reality. Game designers seem relentless in their push for more realistic graphics, often failing to explore other potentials within the medium. There is no reason why games should embrace photorealistic graphics just because they can. Design teams confront realism as a technical challenge, a set of limitations on what they can achieve as opposed to a creative challenge. In other arts, realism is understood as an aesthetic option, one thing the medium can do. In cinema or painting, say, the push towards realism is held in check by a push towards expression or abstraction. The absence of such a counterbalance in games means a gradual narrowing of the visual styles present in games. We would personally welcome games which embraced stylization and exaggeration, which offered us radically different experiences.

'Spacing Out' (November 2004)

Critics often attack games for a perceived lack of good stories. Most games, they argue, boil down to 'save the princess' or 'shoot the demons'.

Leave aside for a moment that there *are* great games with great plots, from the days of Infocom up to *Knights of the Old Republic*. Anyone who reduces a game to its plot does not appreciate the distinctive ways games tell stories through the creation of emotionally compelling spaces.

There is a reason that game guides are called 'walk-throughs'. Walk through the shadowing corridors of *Doom3* or the spritely island landscapes of *Super Mario Sunshine* and space's emotional impact on game experience is obvious. The best designers couple atmospheric design with the challenges and goals which shape our ability to move through these spaces to create mood, rhythm and just plain fun. Great game design owes as much to architecture or dance as to literature.

Recall the feeling of oppression, claustrophobia and chill that accompanied the experience of traversing the underground of the Black Mesa in *Half-Life*. Now, recall that feeling when, after hours of underground seclusion, you first lifted your head into the desert air. Remember the blinding sunlight and hopeful blue skies (too bad that they soon gave way to the desolation of the desert).

As game developers master the building blocks of their medium, we have seen even more subtle uses of space to achieve emotional affect. Take, for example, *Eternal Darkness* for GameCube. The player is Alexandra Roivas, a young woman returning to her family mansion in Rhode Island to investigate her grandfather's gruesome death.

Gramps was involved in something pretty big, involving a plot by 'the Ancients' to use him (and others) to gain power and presumably, reign over the Earth and control the Universe. The player uses his Tome of Eternal Darkness to travel through time to Persian and Cambodian tombs, a French cathedral and the haunted family mansion. Each locale makes an excellent game-space full of awe-inspiring vistas, beautiful light, twisty corridors and secret passageways.

On each return to these locations as a different character, one has a different experience and comes away with vivid new impressions. Consider the cathedral in Amiens, France, which you visit as Anthony, a ninth-century page to Charlemagne, as Paul Luther, a 15th-century monk, and again as Peter Jacob, a reporter in the First World War. In the five centuries that pass between each level, the space is transformed as rooms are closed off, added and otherwise transformed. Players use their knowledge from each previous level to find hidden rooms, locate objects or anticipate obstacles.

The result is an intellectually interesting and emotionally compelling game experience. You first visit the cathedral as Anthony, Charlemagne's page who is trying to save the Holy Roman Empire from being possessed by the Ancients (you fail). When you return as Luther, your goal is to uncover a conspiracy (someone – presumably the Ancients – has framed you for murder). As you try to save your soul and uncover the evil plot, you encounter Anthony's decaying body and possessed spirit. The game taps our memories of Anthony, the cathedral and the Ancients heightening our fear and loathing.

Last, you revisit as the reporter, investigating the mysterious disappearances of soldiers who are being treated in the cathedral, now a field hospital. Having already established the cathedral's terrifying history, the game surrounds the player with fallen and recovering soldiers who are being tortured by the Ancients. Holiness and evil are juxtaposed within the cathedral, creating emotional tension within each era; each successive level plays off the previous one.

Our ability to move through or control game spaces shapes our perception of our characters. In the MMO *Lineage II* for example, players start the game in unique parts of the world, each tailored to that race's backstory and myth. So, orcs are born in a fiery temple, emerging into a stark, isolated land of steep cliffs and dark colours. By the time you reach level 20 you have become a hardened orc. When players finally do leave the orc island, initial encounters with sunlight, different races and even trees and grass are bewildering. By carefully sculpting the environment, the designers of *Lineage II* gently nudge players toward thinking as an orc – mistrustful and suspicious of the outside world.

The emotional work being done by games cannot be reduced to plotlines, anymore than a plot outline of an opera would do justice to the work. *Eternal Darkness* is about more than creepy cults; it's also the story of a cathedral and of the characters who met their fates there. And *Lineage II* shows us that good spatial design can shape and define how each character relates to their surroundings. Until we understand how spatial design shapes our emotional experience, we will not grasp the distinctive aesthetic potential of this medium.

Part II
Space and Time

5
There and Back Again: Reuse, Signifiers and Consistency in Created Game Spaces

Peter Berger

Most games strive to achieve what is called, colloquially, *immersion*. This means that the game world – or virtual space – is so interesting and so detailed that the players forget that they are playing a game. Another word used to describe this phenomenon is *mimesis*. There are games that deliberately transgress against mimesis[1], and games that do so inadvertently, but it is generally accepted that improving mimesis in a game makes it more fun to play (Giner-Sorolla, 1996).[2]

There are many techniques for increasing the immersion of the player in a created virtual space, and improving the mimesis of that space. Four of these stand out as particularly effective: referencing the real world, familiarity and reuse, using signifiers, and maintaining geometric and logical consistency.

The real world, and the virtual-but-real

By 'created virtual space', we refer to the area of play, background, environment and atmosphere that make up the game's virtual location. We describe these spaces in several ways. One method is to describe game spaces as 'purely virtual' (constructed without reference to proper nouns in the player's world), or as 'virtual-but-real'. A space that refers to real or fictional locations that the player has experienced (such as 'New York City' or 'Narnia') is a virtual-but-real space.

A virtual-but-real space, then, is one that leverages real or fictional characters, locations and events that exist outside of the game's narrative, but inside the player's experience. The player's pre-game knowledge allows the game-maker to leverage the player's experiences rather than adding more detail to the story. Players will readily project their knowledge of events.

47

Compare:

Michael rode in to Butler City through the Northern Gate. The
D'jango stadium awaited him. The race would be his; this would be
the year that he would win the D'jango Championship.

With:

Michael Schumacher drove in to Nürburgring via the main gate. The
Nordschleife awaited him. The race would be his; this would be the
year that he would win the Formula 1 Championship.

Or even:

Gaius Junius Clodius drove in to Rome via the Flaminian gate. The
Colosseum awaited him. The race would be his; this would be the
year that he would win the curule aedile's Ludi Romani.

This is not only a question of genre. The psychological effect is strong.
To take a personal example: I enjoy racing games. Racing games activate
my love of cars, and put many vehicles that I can't afford into my hands.
Some racing games are 'tune-up' games: buy cars, and then purchase
aftermarket parts, adjust the shocks, choose the right tyres, and so on, to
improve your car and win races. Two of these games are *Gran Turismo 3*
(2001) for the PlayStation 2 and *Sega GT 2002* for the Xbox.

From a pure driving perspective, *Sega GT 2002* is the better game. The
cars handle more realistically; the cars' response to tuning feels better.
The graphics are better. So are the controls. Yet whenever I want to play a
'tune–up' game, I reach for *Gran Turismo 3*. *Sega GT 2002*, which delivers a
'better' driving experience, sits on my shelf, unplayed. Why?

Because *Gran Turismo 3* has tracks based on real locations, and *Sega GT
2002* does not. When I play *Gran Turismo*, I can drive at Laguna Seca.
I can drive around a real space, a space that existed in my head before
I bought the game. *Sega GT 2002* is full of a number of well-balanced
tracks that allow for great racing. But they aren't virtual-but-real; they
are simply virtual.

So a sense of place can be an important element of games. For some
players, such as myself, this is critical.

Microsoft Games learned this lesson well; *Project Gotham Racing* (2001)
takes place in New York, Tokyo, London and San Francisco. Its sequel
takes place in a range of other cities, including Washington, DC,

Stockholm, Barcelona and Florence. The marketing for the game sold the cities as much as the cars.[3]

The subtlety is that this issue of mapping a video-game model into the player's model has nothing to do with games *qua* games. All games have a set of abstract rules that describe them. All racing video games are, roughly, 'move your piece in a circle, and finish first'. It is the details that provide the elements of fantasy, narrative and drama that separate games from mere exercise. The location doesn't even have to be a real-world location. *Day of Defeat* (2003) is like most other shooters, except that many players have seen the first battle scenes in *Saving Private Ryan* (1998). When these players storm the game's beach in Normandy it points to the memory given to them by the movie, which pointed to a place that, in some sense, doesn't exist any more. This is one example of what separates video games, as a form of mass entertainment, from movies. Filmmakers understand that the location of a movie is as important as the main character. Many game designers haven't yet internalized this.

Creating a virtual-but-real space is one technique for maintaining mimesis. But not every game can or should take place in Times Square. It's unfair to make a space take place in Dayton, Ohio. If you are describing a space that doesn't signify a real-world space, how do you make the player care? How do you coerce the player into thinking of your game space as a 'real' place?

Familiarity and reuse

One of the simplest techniques is reuse: restrict the players to part of your virtual space until they begin to identify with it. This technique is the easiest because it is easily implemented. It is the least trustworthy because it may not fit the game's narrative requirements. If poorly done the player will rebel against boredom.

There are two variants commonly deployed: 'hub and spoke' (typical in platform games), and 'safe house' (common to CRPGs). A place may be both a 'hub' and a 'safe house' at the same time. The difference between the two is one of emotion: a hub is a place that one *transits* to reach more interesting places. A safe house is a place that protects one from the cruel world. Grandma's house in *Legend of Zelda: The Wind Waker* (2003) is a safe house. The hideout in *Grand Theft Auto 3* (2001) is a safe house. The witch's castle in *Banjo-Kazooie* (1998) is sprinkled with hub areas that the player must cross and recross. The cities in *Diablo II* (2000) are both hubs and safe houses.

A safe house can be justified in most narratives without breaking mimesis. Players will familiarize themselves with areas near safe houses unless the designer works to make them boring. Hubs are trickier. They are frequently used as a 'magician's choice' by poor magicians. Consequently, a carelessly designed hub will be treated just as a connection to the next 'good part'. When players can see the wires maintaining the illusion, the result is boredom. If the hub serves no narrative purpose, eliminate it. If the designer must force the players elsewhere, then *really* force them ('Poof! You're in Emerald City now'). Forcing the players to trudge across the world for no narrative reason will earn their contempt.

Provide good reasons for the player to trudge across the virtual city, and it will not be a trudge.

Reuse has some implementation benefits. Reusing program assets saves on development cost. It establishes a game location as a location for purposes of later licensing, or for use in sequels. Reuse as a trope matters most to the designer who is pressed for time. But this more than any other technique requires the most effort on the part of the player. By using reuse to establish immersion, the designer is balancing the provision of a stable area of play against the risk of boredom.

Reuse of created game spaces is risky because most games are already limited in certain ways. Most modern games have a few core game mechanics that are static throughout the game. This means the ludic challenge of the game designer is developing 30 seconds of fun game-play, and then stringing those 30 seconds together for 10 or 15 hours to hold the player's interest.[4] Familiarity is troublesome because it will call attention to this aspect, rather than conceal it. When possible, designers should use additional techniques to enhance immersion rather than solely relying on familiarity and reuse.

Internal signifiers

The player is part of a narrative. Designers may be able to use signifiers to build the player's map of the created game space before depicting that space. By signifiers we mean text, maps, road signs, graffiti, conversations or any technique that allows the designer to suggest part of the game world to a player before she arrives.

This can be used to provide not only knowledge of geography, but of objectives. 'Oh!' said the Princess, 'the Eastern Forest is overrun with wolves!' 'Right,' thinks the player. 'East, forest, wolves. Got it'. One line advances the game's narrative and provides the world's topology.

Signifiers do not have to be straightforward. They are a mechanism to manage expectations on the part of the player. This means they can subvert expectations as well. The *Silent Hill* games do this magnificently. Players can obtain maps of the areas they are in. The maps are rough sketches – gas-station maps, fire-escape diagrams and the like. The player reads the map and says 'Oh, I can travel west on King Avenue and reach the park'. The protagonist heads west and encounters a water-main break – the road is closed. When he discovers this, the game marks King Avenue with a red line. The player has been given topological information about the space, organically, in a way that makes narrative sense: mimesis is preserved.

As clues and obstacles are encountered, the player's map automatic-ally updates with notes, circles, jagged lines. The player quickly learns that the unannotated map is untrustworthy – the straightest line in *Silent Hill* (1999) is never the correct solution – but it is all he has got. The liminal urge motivating players is to unearth the narrative. The ludic urge motivating players is to kill zombies and solve puzzles. But the subconscious urge, I would propose, is the desire to fill in the map.

One reason these maps work is that they mimic a believable element of the physical world: a paper map and a red pencil. Some games provide 'radar' maps giving the player a precise view of the game world.[5] While functional, these tools can hurt the game more than they help it. They can break mimesis, and distract the player from the meticulously created game world. If players are distracted and not paying attention to the vir-tual space, then they're not internalizing it, and thus it loses impact.

To reduce this to a rule of thumb: never show the player the same map that the game uses internally. Providing a separate map allows the designer to make choices about what to include or omit, and provides the oppor-tunity for deploying a map that disagrees with the game's internal map. Disagreement between a signifier and the signified tends to increase the players' concentration on the environment, as they try to work out why the contradiction happened. Obviously, how you present that contradic-tion will have narrative impact – a trusted ally maliciously lying to play-ers has one set of narrative consequences, whereas allies who are simply mistaken has another. And if a signifier, be it text, map or a character is always wrong all the time, the player will simply discard or ignore it.

Consistency

Consistency gives the sense that the world works by a set of rules.

In Sony's *Ico*, the protagonist is a young boy trapped in an ancient, decrepit castle. The edifice itself is more than the environment the prot-agonist moves in. The edifice is his true opponent. The castle you are try-ing to escape from is the nemesis. The monsters you encounter, the Witch-Queen who imprisons you, are not half as impressive and oppressive as the place you are trapped in. The monsters are shadowy figures, easily dispatched with the thwap of a stick. The castle is bigger than you. The castle is older than you. The castle has seen thousands of boys like you come, and none have ever left. Hit the walls with your stick. They will not fall.

One thing that makes this work is the internal consistency of the game-space, which is intimately tied to its visual style. It's not just that the castle is a consistent mappable space which obeys the laws of geometry and physics – it's that this is shown to you, deliberately, time and time again. You leave a room and step out on to a balcony. In the distance is a hexagonal tower. You struggle through three or four more rooms and find another balcony; that tower is closer now. Forty-five minutes later, you are at the base of the tower. Climb it, and look back, and you can see the balconies you were on early. That is where I was. This is where I am. Over there is where I'm going.

The introductory montage in the classic game *Half-Life* (1998) is a great example of consistency and expectation-setting. The pace is positively glacial. Nothing happens; you are in a tramcar which trundles through the Black Mesa complex, carrying you into your lab at the heart of it. What it accomplishes, though, is that it gives you the sense that this is a fully realized place, with its own geography. On your way out through the shattered complex you will pass some of the places seen on the way in. This self-reinforcement enhances the experience. (It's one reason why the 'Xen' levels near the finale of *Half-Life* are so weak – the player goes from exploring a consistent, thoroughly realized world into a fantastic world, a fantastic world that looks like it could have been lifted straight out of Super Mario World, complete with platforms that move for no discernible reason other than to let the player reach the boss monster.)

The tradition of attempting to provide a physically consistent world is as old as computer games, with even the original *Colossal Cave* text adventure presenting the player with an internally consistent (although in places confusing) plan, complete with foreshadowing of places yet to be visited (think of its famous 'mirror room'). A consistent, believable model of physical space is a precursor requirement to the location itself having a meaningful impact on a game. There is no guarantee that a consistent, believable model will be interesting, but a model into which no thought has gone will be nothing more than window dressing. This is, obviously, one of those 'know the rules to break them' situations – obviously a dream world might have different rules than the real one, likewise the hallucinogenic flightscapes of *Rez* (2001). But if you simply throw together a virtual space without a vision of where the player belongs in it, the players will be able to detect your laziness on an almost subconscious level.

Consistent worlds also provide the game designer with the choice to break the rules in a way that will have ludological, narrative or psychological impact. Continuing with our *Colossal Cave* example, parts of the text adventure consist of a maze. Whereas most of the game provides rich, detailed descriptions allowing the player to orient herself, once in the maze the only description is: 'You are in a maze of twisty little passages, all alike.'

Subverting consistency in a controlled way provides the designer with the opportunity to confront the player with a 'game within a game'. The labyrinth, in particular, has been a staple of text adventures since their earliest incarnations. Labyrinths appear in many Infocom games, including *Zork* (1980). One interesting recent trend in interactive fiction design is the ascendancy of the 'anti-maze'. An anti-maze is a portion of the virtual game space which superficially appears to be a maze, but which is not solvable through brute-force search: it requires the player to solve a puzzle to escape.

Perhaps one of the best examples of an anti-maze is in Andrew Plotkin's *Hunter, In Darkness*,[6] an award-winning tribute to the primitive maze game *Hunt the Wumpus* (1976). In *Hunter*, towards the end of the game, the player is tracking the wounded wumpus and finds herself in a room of interconnected caves. Each cave has marvellously detailed descriptions, indicating the colour of the nearby rocks, the character of the lights, and an indication that the player can move in any direction. No matter how carefully this space is mapped or traversed, the player can never escape; the game will simply randomly generate more rooms. The anti-maze solution is for the player to stand still and wait, and after a time a group of bats – whom the player earlier learned are attracted to the scent of blood – will converge around an exit. The bats are showing the player the route followed by the injured Wumpus. An anti-maze with similar (though less intricate) mechanics appeared in *Zelda: The Wind Waker*. That mazes and anti-mazes have narrative impact is made possible by the consistency of the world that exists *outside* of them. One cannot see the figure if there is no ground.

Consistency can also simply be used for managing or subverting narrative expectations. In Silicon Knights' 2002 survival horror game *Eternal Darkness*, the created game-space generally operates under consistent (if malign) rules. In addition to the 'health meter' that is standard in the genre, the player has a 'sanity meter' as well. Each time the players' characters confront an unearthly horror, they lose some of their sanity. If the character loses too much sanity, the consistency of the world begins to fall apart. The camera view will tilt. The walls will bleed. The game will also, in this mode, begin to break mimesis in particularly clever ways: it will appear to turn the volume down on your TV, for instance, or tell you that it has deleted your saved games, or made your game console crash. Normally, such antics would break mimesis and make the game less effective. Perhaps because of the subject matter, the overall effect in *Eternal Darkness* is to decrease the players' confidence in the dividing line between their characters and themselves. It is supremely unsettling, and an example of breaking the rules to great effect.

Such transformative moments in a game are only made possible by the presence of consistency. If the designer does not take the time to establish the consistency of the game-space, then the impact of 'violating the rules' is dissipated, because there are in fact no rules to be violated yet. This creates a minor conundrum: to tell a truly radical story – at least, one with any narrative impact, and not mere formalist indulgence – the beginning of the story must be told in a reactionary way.[7]

Computer and video games go beyond mere ludic experiences, but include elements of fantasy play, character identification and narrative immersion. Deeper immersion generally makes for a better game. Properly deployed, references to real or fictional worlds within the player's experience, careful reuse of in-game locations, aggressive and creative use of believable and mimetic in-game signifiers, and a consistent approach to architecture and design will all serve to enhance the immersive qualities of a game.

Intentionally breaking mimesis should be avoided unless there is a strong narrative reason to do so. In order to effectively break the rules in order to create a narrative impact, a game is obligated to first establish and communicate the rules that are being broken.

Notes

1. Often in the form of 'breaking the fourth wall' between the player and his on-screen representation. For example, in the 2005 remake of *The Bard's Tale*, in-game characters tell the protagonist to 'press the Y button', which bewilders him.
2. Giner-Sorolla discusses some of the common tropes and mechanics surrounding text adventures (the prevalence of mathematical puzzles, for example, or the ability to save and load the game) and discusses how these damage the mimetic ability of the medium. This essay later inspired Adam Thornton's IF game *Sins Against Mimesis*.
3. A representative Microsoft press release text for *Project Gotham Racing 2* reads as follows: '*Project Gotham Racing 2* (Microsoft Game Studios, Bizarre Creations

Ltd) is the ultimate test of racing skill, style and daring, rewarding drivers not only for how fast they drive but for how they drive fast. Drivers will earn kudos and gain recognition for cornering on two wheels around the Sears Tower in Chicago, power sliding through the ancient streets of Florence, Italy, or manoeuvring along the racing line through the slick streets of Edinburgh, Scotland, all while racing some of the most exotic high-performance vehicles available. Players also can challenge friends and other gamers online utilizing the Xbox online game service or through multiplayer System Link.' In other words, in promoting their car-racing game, when space was at a premium and only the minimal message could be delivered, Microsoft marketing chose to mention no cars, but spent approximately half of their message discussion the virtual-but-real locations in the game (Microsoft 2003).

4. *Halo 2* lead designer Jamie Griesemer said 'In *Halo 1* there were maybe 30 seconds of fun that happened over and over and over and over again. So if you can get 30 seconds of fun you can pretty much stretch that out to be an entire game. Encountering a bunch of guys, a melée attacking one of them before they were aware, throwing a grenade into a group of other guys, and then cleaning up the stragglers before they could surround you. And so, you can have all the great graphics, and all the different characters, and lots of different weapons with amazing effects but if you don't nail that 30 seconds, you're not going to have a great game' (Behind the Scenes, 2004).

5. The games in the *Grand Theft Auto* series, for example, provide an overview map showing the player not only his current position, but the location of his next mission and all of his contacts and safe houses.

6. *Hunter, In Darkness* (1999) is also notable for Plotkin's use of traditional narrative techniques to breathe life in to a sparse setting. *Hunt the Wumpus* (1976) took place in an empty cave. The cave was an undirected graph of rooms with the player, the wumpus and a bat. The rooms themselves were featureless. In Plotkin's retelling, the cave comes alive with detail and menace. The effect is a first-rate demonstration of the importance of setting and environmental detail to a narrative.

7. That the method of telling a story is traditional does not require that the plot, setting or characters be traditional. In Sega's cyberspace rhythm and shooting game *Rez* (2001) the narrative takes place entirely within a 'virtual reality' interface to a computer system. Most of the images and settings in the game are completely formalist and abstract. Despite this, the settings change in degrees, and even though the world is *fantastic*, and although many elements of it change over time, there are still consistent aspects that the player can (and indeed, must) hold on to.

6
Another Bricolage in the Wall: Deleuze and Teenage Alienation

Jeffrey P. Cain

'What child is there that, coming to a play, and seeing *Thebes* written in great letters upon an old door, doth believe that it is Thebes?'

Sir Philip Sidney, *Defence of Poesy* (1595)

After the gruesome murders at Columbine High School in 1999, the news media began to focus considerable attention on computer video games as a possible cause for Dylan Klebold and Eric Harris's violently antisocial behaviour. Perhaps the most influential statement of the premise that video games serve as training tools for killers came from one Lt Col Dave Grossman, co-author (with Gloria DeGaetano) of *Stop Teaching Our Kids to Kill: A Call to Action Against TV, Movie, and Video Game Violence* (1999). Colonel Grossman, who routinely offers presentations on a subject that he has named 'killology', appeared on the 24-hour news channels and before Congress to ascribe to games such as *Doom* the power to make murderers, sadists, perverts and suicides out of otherwise normal teenagers (Grossman and DeGaetano, 2006). This thesis has gained in journalistic and talkshow popularity ever since Columbine. *Doom* (1993), *Quake* (1996), and numerous other games are widely considered to engender an addiction to violent behaviour.

However, it is not just inherently violent themes that have been prone to charges of promoting disaffection among young people. In 2001 a young man named Shawn Woolley committed suicide, ostensibly in connection with events that had occurred deep inside the hugely successful online role-playing game *EverQuest*. His mother, Elizabeth Woolley, hired a well-known personal-injury lawyer, Jack Thompson, and threatened to sue Sony Online Entertainment and its subsidiary, Verant, which is the company that created and now produces *EverQuest* (usually called *EQ* by gamers). She publicly accused Verant of literally killing her son with images. Shawn Woolley, she argued, was epileptic and had suffered seizures that were caused by staring into a computer monitor for hours. She blamed events that occurred inside the game for his suicide.[1]

It might be said that both Colonel Grossman and Mrs Woolley were engaged in the same futile mission: an attempt to contain the machinic assemblages with which a given (presumably innocent) human body connects to produce image-texts and virtual events. Grossman and Woolley's contention that a static wholeness pre-exists a teenager's connection with gaming obscures the phenomena at hand. Nevertheless, their instincts were right in one sense: it is a serious mistake to underestimate video gaming by regarding it from the viewpoint of naïve representational thinking, as a mere series of images and sounds that function as signifiers of a fictional world.

The safe suburban values compromised by video games lie in what Gilles Deleuze and Félix Guattari would call the 'in-between'. Suburbia can be thought of as a series of conjoined strata that fold into the city on the one hand and unfold into the rural on the other. Suburbia is the *'and ... and'* in the continuous and recursive topographical origami of rural and urban.[2] Therefore, the publicity that followed Columbine focused the anxieties of suburban parents, who view drug addiction as the most immediate social threat to their children. Here was a new horror: a drug that was not a substance but a machinic assemblage that replaced apparently stable sociogenic values with a radically unstable and open-ended process of becoming 'other'. Games such as *Quake*, *EQ*, and their many successors produced effects that, to adults, simulated drug abuse. For years, parents and schoolteachers had blamed television for eroding morality and intellect, but video gaming was even worse: it was like television on crack. Gamers who revelled in the authorities' dismay sometimes called it 'Evercrack'.

Computer game antipathy involves a model of addi(c)tion as a kind of *addition*, a more or less substance-free desiring production that deploys itself into the virtual in order to create and open up a multiplicity of new connections, manners, styles, speeds and vectors of reality.[3] The purpose of this essay is to extend the role that Deleuze and Guattari's thought models play in the academic discourse on gaming. Many different Deleuzean concepts might serve as starting points, but the most relevant method is micro-analysis, an as-yet incompletely defined act of empiricism that seeks the granular levels of otherwise imperceptible events within a machinic assemblage.[4] Deleuzean concepts afford vantage points from which to consider the physical, narrative and virtual spaces in which video gaming transpires. Moreover, the game worlds offer perceptual positions from which to generate strange views of ordinary reality. As an illustration, I will offer a critique of *EverQuest (EQ)*, a game I have chosen partly because of its compelling influence on the world of gaming, and partly because I have actually played it myself.[5]

Prior to the fold: exchange and 'Evercrack'

The Deleuzean concept of the fold serves to organize perception of a world subject to continuous and dynamic becoming. It is the fold that gives *EQ* the appearance of interiority and exteriority even though the difference is illusory. In order to perceive *EQ*, we need to follow the process of folding and unfolding with which the game world articulates and records its own flows of becoming and difference. Each new player ventures upon the recording surface of the game in the midst of the action, although not in the traditional Horatian sense of *in medias res*, since there is no singular or overarching narrative. Nonetheless, certain connections and conditions pre-exist the fold that will wrap the player inside the game.[6] One of these is the principle of exchange.

There is always already a virtual economy of addiction, a relationship of the addict's percepts and affects to the street value of whatever it takes to reproduce or chase the high.[7] This principle is nowhere more evident than in *EQ*, which is one of the oldest and most influential massively multiplayer online role-playing game (MMORPG) in existence. Thousands of people play *EQ* simultaneously online and every other MMORPG is in some sense derivative of it. The *EQ* game world of Norrath has its own currency, of which the basic unit is the 'plat', short for 'platinum coin'. High-level players amass huge sums of plat by fighting and killing monsters, raiding treasuries, buying and selling weapons and completing quests.

However, it takes enormous quantities of time and skill to do so, and many newbies want expensive armour and weapons right away. So plat is traded on eBay and other websites, and can be bought for US dollars and other currencies. A player can kill a dragon inside the game and earn 1000 plat, which can then easily be converted into real dollars by selling it to a plat broker. The last time I checked, one million plat – enough money to buy anything a player could want in *Everquest* – cost about $500. An Indiana University economist, Edward Castronova, analyzed the overall value of the *Everquest* economy and found at the time that one plat was worth roughly $.01 in 'real world' value, a bit more at the time than the Japanese yen or the Italian lira. Castronova discovered that the average player generates about 319 plat per hour. That's $3.42 an hour, more than the daily wage rates in some third-world countries. *EQ's* yearly GNP was $2266 per capita, making it the 77th-richest country in the word, ahead of India, Bulgaria and China (Thompson, n.d.).[8]

Thus the exchange model in *EQ* is not contained inside the game. Indeed, the game's exchange network draws power from the everyday world though connections that are cultural, political, economic and biological. Progress in the game requires biological real time spent online, and thus withdrawn from work, family and school. Real money has use-value only insofar as it can help the addict chase the high. For the delayed gratification of the work ethic, *EQ* substitutes the experience that Deleuze and Guattari postulate as the basis of all drugs, 'a whole rhizomatic labour of perception, the moment when desire and perception meld' (Deleuze, 1987, p. 283). 'RL' is the *EQ* gamer term for 'Real Life', that strange and boring place where players go when not in-game. Obviously, the rich labour of desire and perception in *EQ* is more attractive and fulfilling than is working in a pizza parlour, which it seems was the last steady job Shawn Woolley held before his death (Winter, 2002).

Most *EQ* players would reluctantly admit that RL is important, but others prefer not to talk about it at all. Elizabeth Woolley and Colonel Grossman thought that there is a clear dividing line or difference between 'real reality' and 'game reality', but this conviction is precisely what defeated them. Deleuzean thought inverts the structuralist notion that difference is something imposed or constructed upon a separate and otherwise chaotic reality (Colebrook, 2002, pp. 38–40). It might even be said that RL is itself subject to a 'becoming-the-game', or a 'becoming-*EQ*' that operates via duration and immanence, in which daily life pales into a mere interlude between gaming sessions. Players who can collect unemployment checks or find an undemanding job can spend most of a life at home in Norrath and simply commute to RL.

Although *EQ* is not a substance, it comprises flows of energy, passage and speed, which connect with the body and can be abused. Peta Malins points out that 'while Deleuze and Guattari do propose one way of accessing the positive desiring lines of drug use without plunging toward a black hole, even this is only by skipping the substance itself' (Malins, 2004, p. 94). As Deleuze and Guattari remark, drugs are neither essentially nor necessarily substances; instead, drugs are more like catalysts that alter, divert or striate flows of speed and perception: 'What allows us to describe an overall drug assemblage in spite of the differences between drugs is a line of perceptive causality that makes it so that (1) the imperceptible is perceived; (2) perception is molecular; (3) desire directly invests the perception and the perceived' (Guattari, 1987, p. 282).

Speed is a percept of the relation between the spatial and the durational. The new perceptions are not generalized impositions of difference upon a prior and chaotic universe, but a *less*-generalized or more focused view of already-existing pure and positive difference (Colebrook, 2002, p. 38). Thus Deleuze and Guattari (1987, p. 283) refer to the 'mad speeds of drugs and the extraordinary post-high slownesses'. Any addict, including the *EQ* addict, experiences this slowness simply as (real) time that must be endured until the next high. This connection intensifies as the player's attention begins to focus on the granular game events. As we shall see, a rhizomatic web of associations obtains between and among the game, the players and such approximately centripetal Deleuzean concepts as desiring production and the sundry kinds of becoming. None of these connections is overarching, totalizing, naturalized or molar. Instead, the game encompasses an immanent series of smoothly perceptible yet carefully segmented recording surfaces, both visual and mathematical; examples include plat accumulation, experience points (XP), and ability ratings for everything from casting spells to tolerance for alcohol, all of which are calculated to keep players in the game by recording and quantifying productive desire. Each battle allows the game body to accumulate XP, essentially a mathematical expression of how much immanence and becoming a given character has withstood.

The main activity in *EQ* involves fighting MOBs ('mobile objects' or monsters) that guard a specific area: a cave, castle or dungeon. The fights thus precipitate a continuous process of de- and re-territorialization (Deleuze and Guattari, 1987, pp. 508–10; Colebrook, 2002, pp. xxii–xxii). Some time after being vanquished, monsters regenerate themselves, thereby reclaiming their territory until the next group offers battle. The players are thus engaged in a nomadic project of deterritorialization; they take territory and hold it only for as long as is necessary to remove whatever prize, treasure or magical object it contains. Eventually the players leave, but they gain tactical knowledge that will help on the next trip to that same area and have thus partially reterritorialized that part of the map. However, a rhizomatic map is never a participant in the subject-object split, never a separate tracing of a prior and static reality. Instead, the rhizomatic map offers a one-to-one correspondence with reality through connections that emerge as the map expands asymmetrically. Thus *EQ* is itself an ever-expanding deterritorialization of RL and in the end, the game addict pursues the line of flight as thoroughly and quickly as possible. In a sense, addiction is deterritorialization without the hope of reterritorialization, at least along the line of flight from the quotidian.

Every *EQ* character derives from an archetype called the avatar, which lends certain features and abilities to the players' game-bodies. Elves, for example, have pointy ears and a high intelligence number. Ogres and Barbarians are extremely strong, and Druids can find food in places where other players perceive only grass and trees. All the characters/ avatars are born in a home area, but are soon forced to become nomadic in order to find bigger MOBs to fight and territories to plunder. All Barbarians, for example, are born in an icy northern city named Halas. Once a fledgling Barbarian has hunted for a while in Halas, it becomes necessary to explore further in order to gain experience points (XP) by finding more difficult battles. Gaining XP raises one's level, and the higher the level, the greater the ability to fight, heal, flee or cast spells. The object of the game is therefore 'levelling up'. Each gain in XP creates new connections with both the game and other players. The game space comprises a multiplicity of unfoldings, affects and percepts. For a player accustomed to the thought models of literary theory, the experience resembles wandering around in a novel, except that there is no closure or ending. Every dénouement is a new exposition, and climactic con- flict, in the form of battle, is an hourly experience. *EQ* has no winner; the experience is infinitely open-ended and the game creators regularly deliver new levels and adventures. Playing for XP is a way of quantifying immanence other than that which the player knows in RL.

Folding in: *EQ* flees the dream

The master trope of naïve moralism, called the 'American Dream', fiction- alizes and transcends all societal markers, including race, gender and class: 'You can attain anything you want in life, *if only you work hard enough*'. The Dream insists on personal labour as the sole means of pro- ducing not just wealth and prestige, but identity itself. The Dream mani- fests itself as a terrifying juggernaut of representational thinking, fuelled by the imposition of negative difference upon an ostensibly prior and chaotic universe. To a devoted *EQ* player, the Dream is more frightening than any monster or dragon ever conceived in Norrath. Only a suicidal fool would go back to RL to face the Dream.

The Dream defines desire as *lack*. Whatever social position those sub- jugated by the Dream occupy, they keenly feel it to be less desirable than some other and higher level of material, political or spiritual attainment (Holland, 1996, pp. 240–5; Parnet and Deleuze, 1987, pp. 77–123; Colebrook, 2002, pp. 16–18). The Dream is an advertisement addressed to consumers of a product: identity. Each time a True Believer in the

Dream pays the price – financial, psychological, physical, political and spiritual – for a new level of identity ownership, the Dream recedes and gestures toward an even higher state of being. Hence the Dream condemns the consumers of identity to a perpetual state of lack, loss and emptiness.

Video gaming, by contrast, celebrates desiring production; it replaces the gamer's original self with a more fluid version of the ego, the self as pure *becoming*. Or rather, the game facilitates the perception of a pure becoming that has always been immanent. The only limit to the multiplicity of connections between player and video game involves the now-familiar Deleuzean concept of the Body without Organs (BwO) (Malins, 2004, p. 88; Deleuze and Guattari, 1983, pp. 9–16). *EQ* avatars offer gamers a multiplicity of capabilities that attach to the BwO, which operates on a plane of desiring imagination linking the gamers and game creators. The resultant game-body gradually becomes prior to the RL body; this shift occurs as the BwO accustoms itself to passing from game to RL and back again. The cycle of passage and return is recursive but not concentric, because the rhizome offers no centre from which to depart.

The language of the Dream trope is logocentric, sophistic, foundational, representational, metaphorical, rhetorical, totalizing, naturalizing and self-perpetuating. Words serve the Dream as signifiers useful for extending influence and glorifying the meta-narrative of success. *EQ* also uses language in various telling ways, but its basic grammar is neither rhetorical nor logocentric; the unit of meaning is molecular rather than molar. The most important and sought-after words in *EQ* do not in fact possess any traditional semantic value. They embody and deploy the word as pure machinic function: the *password*. It is only through passwords that the game-body may enter and explore the fold that is EQ.

Inside the fold: becoming the game-body

Deleuze once remarked that 'each individual, body and soul, possesses infinity of parts which belong to him in a more or less complex relationship' (Deleuze and Parnet, 1987, p. 59). Inside the fold of *EQ*, the player's game-body employs perception and sensation to focus the game's reality. As in RL, gamers may perceive or feel that their *EQ* character experiences spirituality; thinking of the game-body as in possession of a soul helps visualize game addiction. Once *EQ* prevails as the home of mind and spirit, the world of RL and the individual human body begins to fade. Time contracts; the player spends 11 hours in the game but hardly notices it. *EQ*'s creators at Verant appear briefly in the role of first movers, but it quickly becomes apparent that they too are part of the game's immanence: their art is to transform flow and difference into images and sounds. The borders of *EQ* constantly materialize in virtual space and rush away from the players. The rhizome grows; it maps reality as it expands, its connections vary infinitely with no centre. The becomings of the rhizome and the becomings of the player are one and the same thing, even though the player can never perceive all that is immanent. This process results from the mind and soul's functions as active entities of perception. As such, they are steadily refocused on the game, as opposed to RL.

Inside the fold, the game-body functions much as does an organism in the traditional account of evolution. Thus the players encounter challenges to their continued becoming and either surmount these challenges or die in the attempt. Since it is a postmodern death, however, it can be repeated as many times as necessary. Thus one player-avatar is the cleric, who can raise characters from the dead. A typical *EQ* session sees the game-body acting to deterritorialize some part of the game's topography. *EQ* was originally designed to encourage the players to work together in groups, so many of the monsters to be fought are far too powerful for defeat at the hands of a single player. For important battles, such as the finish of an epic quest, 30 or more players may be required to win, and planning for the assault will start days before the team goes into action.

Once a strategy is chosen and enough players with the correct skills have been recruited, the attack begins and tactical prowess is highly valued. Warrior-class players, for example, must be adept at the skill of 'pulling': luring the MOBs into a position where they must fight without the support of other monsters of their own ilk. The warrior then holds the MOBs at bay while everyone else ambushes it. These tactics are repeated in almost every new area, but as a player's skill level rises and the tactical environment becomes less forgiving, precise teamwork becomes imperative. One error can bring death to all the players involved in a fight. All players seek treasure and XP, but they also need respect from their comrades; too many mistakes in battle and invitations to raiding parties will dry up. Thus the game-body thrives in a rich and adventurous cultural context.

Folding out: the fate of the game-body

EQ might be seen as form of bricolage, a momentary artistic assemblage of images, sounds, immanence, difference, consistency, resonance and variation. Because Deleuze always proceeds from immanence, he does not see the image as a traditional linguistic sign that is then available for interpretation by a transcendent or objective viewer. In *EQ*, the game images are an extension of the game-player. Or to put it another way, the image is not a signifier or representation of the body; instead, the image is itself the body. It might be objected that an *EQ* player sitting at the keyboard still intuits a distinction or split between the RL body and the image body used to travel through the game world. However, the person making such an objection simply has not played *EQ* long enough. *EQ* constantly challenges the fabric of RL and inverts, disrupts or re-channels everything, including space-time perceptions.

The BwO's most important contribution to this process is its diachronic becoming. A player easily begins to imagine the character as an extension or simulation of the organic body. The BwO then challenges all the most treasured assumptions and connections of the player's organism, including the belief that this specific group of organs already belongs together. As every philosopher since Socrates has remarked, the organic body is a major problem, always trying to hang together until the final moment, when only its dissolution promises hope of a reality beyond its myopic organ-*ism*. The BwO, however, allows for a process of 'becoming other' that does not involve a traditional subject-object split. Instead, the BwO opens itself to a series of formerly imperceptible percepts and affects. To the dedicated gamer, the extension of the BwO into the virtual world defines pleasure. To live with a large part of one's becoming always inside the game of *EQ* is to pursue a line of flight from the Dream, to deterritorialize the officially sanctioned percepts and affects that inform middle-class complacency.

Is addiction the only viable path for some people? The Dream is a manifestation of suburban and bourgeois anxiety. The shooters at Columbine were outcasts from the Dream, bullied by other students who were in complete possession of the school's culture of team sports, cheerleading and other wholesome activities. Why couldn't the Columbine shooters and Woolley return to normalcy? Because 'normalcy' makes outlandish and completely absurd claims: there exists a foundational 'real world' that we can all observe via common sense, identity is determined by organism, the goal of life is organic reproduction, the means to continued existence is commercial labour, history is linear, differentiation is structural and semiotic, all knowledge proceeds by taxonomy, you can only be one person at a time, combat is a last resort, there is no such thing as magic, and time (*pace* relativity) is constant.

From the gamer's outlook, nothing could be further from the truth: identity is multiple; magic makes space and time contract, expand, fold, freeze or burn. There are no shopping malls or football teams in *EQ*, just nomadic aggregations of players acting on the most primitive of social contracts: glory in battle. Why would anyone want to return to RL after experiencing a life of imperceptible percepts, lightning speed, gratifying affects and fascinating danger? Shawn Woolley's *EQ* companions, whatever they thought of his avatar-character or his game-body, must have understood this point perfectly. They held a memorial service for him, but they did not meet at some convenient RL location: they assembled to do so inside the game.

Notes

1. For Shawn Woolley, see Marks (2003, pp. 21–34); Frankel (2002); Miller and Winter (2002).
2. For the *between* as multiplicity of inseparable relations see Deleuze and Parnet (1987, p. vii). For the universe as origami see Deleuze (1993, pp. 3–26). Much of my general approach to Deleuze (but none of my error) is indebted to Claire Colebrook's many publications, especially *Understanding Deleuze* (2002). I am similarly obligated to Peta Malins's 'Machinic Assemblages: Deleuze, Guattari and an Ethico-Aesthetics of Drug Use' (2004). Also, I wish to thank Denise Neuhaus for her expert editorial help.
3. For *desiring production* and *desiring machines,* see Deleuze and Guattari (1983, pp. 1–42).
4. For variations on 'micro-analysis' see Deleuze (2002).
5. Before quitting *EQ* in 2004, I had played my character, a Barbarian Warrior named Barudil, to level 54. Barudil lived a rare life of speed, strength, danger, humor, love and pleasure. To my former comrades of The Well of Souls: *ave atque vale!*
6. For a more psychological reading of subject development and the fold, see Semetsky (2004, pp. 212–13).
7. For the nuances of percept and affect see Deleuze and Guattari (1994, pp. 163–200).
8. See also Jordan (2003). For the original paper, see Castronova (2001). For the market value of cyberspace avatar-bodies see Castronova (2004).

Part III
War and Violence

7
Programming Violence: Language and the Making of Interactive Media

Claudia Herbst

While not all games feature extreme forms of violence, many of the most successful releases involve particularly brutal and bloody forms of game-play, alarming parents, politicians and theorists alike. Contributing to a rich body of inquiry, this essay investigates the role of computer languages – code – in the creation of gaming content, its structures and narratives. Arguably, computer languages encapsulate interactive media products similarly to how a screenplay encapsulates a movie plot: language, structure and content are inseparably intertwined. When a programmer writes the interactivity for a computer game, he or she not only defines a game's functionality but also its narrative structure and, to a degree, the content of a game. Technology's artificial languages – code – differ from so-called natural languages and it is precisely because of this difference that gaming narrative diverges from traditional narrative forms. The trendsetting games *Spacewar* (1962) and *Doom* (1993) offer examples of how code informs narrative.

Writing games

Computer games are not written in natural languages but are experienced based on the interactivity provided by a programmer. A programmer may perceive of the role of computer languages in the creation of interactive content as painfully obvious. To most, however, the fact that computer languages play such an important role in the creation of interactive content is concealed by the interface. One of the first computer games, written by programmer Steve Russell in the early 1960s, is *Spacewar*. Russell's creation was the first game to have been written with a screen and a typewriter instead of punch cards, and it quickly made gaming history and created its own genre, shooter games. Russell's work process is described as follows:

> He then set about making the shapes of the two rocket ships: both were classic cartoon rockets, pointed at the top and blessed with a set of fins at the bottom. To distinguish them from each other, he made one chubby and cigar-shaped, with a bulge in the middle, while the second he shaped like a thin tube. Russell used the sine and cosine routines to figure out how to move the shapes in different directions. Then he wrote a subroutine to shoot a 'torpedo' (a dot) from the rocket nose with a switch on the computer (Levy, 1984, p. 48).

The description of the creation of *Spacewar* combines references to mathematical functions (sine and cosine routines) and the making of objects such as rockets as though they were physically built. Yet, none of the elements of the game, the rockets and torpedoes, exist in real space; they are virtual manifestations of code. The account continues with a description that, semantically, is more accurate: 'Then he wrote a subroutine to shoot a "torpedo" (dot) from the rocket nose with a switch on the computer' (Levy, 1984, p. 15). That the words 'making' and 'writing' are interchangeable in the description of Russell creating *Spacewar* illustrates the essential role computer languages play in computer games.

Like *Spacewar*, the game *Doom* was designed and created by its programmers, a small group of twentysomethings who had formed the gaming company id Software. It was the programmers who conjured up the elements of the game *Doom*, such as its architecture and characters, as well as its functionality. In the chronicles of *Doom* (Kushner, 2003), when a team member suggests that the story of *Doom* should be written down (in natural languages), one of the lead programmers objects. *Doom* didn't need a 'back-story' (Kushner, 2003, p. 132). The project had been well on its way and the lead programmer knew that he had been 'writing' the game all along – in computer language. In the case of *Doom*, writing the story and programming the game was one and the same process. After the game became a huge success, one of its lead programmers proudly took to wearing a self-designed T-shirt that featured the *Doom* logo and the phrase, 'Wrote it' (Kushner, 2003, p. 172). The role of computer language in the making of a game is particularly evident in the annals of *Spacewar* or *Doom*. Their respective stories illustrate the central role that computer language plays in interactive media. In a computer game, computer language – code – is arguably not one, but *the* crucial element, the glue that holds all the other elements together, the lifeline that provides functionality.

Programming is a skill through which a game's functionality is encapsulated. The interconnectedness between code and interactive games becomes evident when we consider that without computer languages, interactive media would not exist. Yet, there arguably exists a deeper connection between code and content – one that manifests itself in the dominating type of gaming content, which includes repetitive and ruthless violence.

Mayhem and enactment

Most seasoned game players will confirm that interactive media are capable of inducing a real sense of danger and fear. Some of the descriptions of game advertisements and reviews may read as though the violence contained in them is so exaggerated that it will not be taken seriously, such as when, for example, a cartoon character is hit by a boulder. Nothing could be further from the truth. Commenting on the effects of interactive experiences, Simon Penny (Wardrip-Fruin and Harrigan, 2004, pp. 79–80) notes, 'The user is trained in the enaction of behaviors in response to images, and images appear in response to behaviors in the same way that a pilot is trained in a flight simulator'.

The implications of enactment of violent behavior are far-reaching. *Doom*, along with the equally violent *Duke Nukem 3D* (1996), has been linked to the Columbine High School massacre that took place in Littleton, Colorado, in 1999. In a wild shooting spree that resembled the carnage depicted in those games, two boys shot 12 students and a teacher, and then turned the guns on themselves. The game *Counter-Strike* (2000) has been linked to a school shooting in Erfurt, Germany, in 2002, where a student shot 13 teachers, two students and a police officer, before he, too, turned the gun on himself. Joshua Goldstein writes:

> In 'shooter' games the screen shows what the player would see as he blasts away at realistic people (creating realistic wounds, such as severed heads). A rating system is supposed to keep these 'whack-and-hack' games away from young children, but does so imperfectly. Empirical research has not shown compelling evidence that playing violent video games make children – rather boys – behave more violently. These games are relatively new, however, so not many studies have analyzed this connection (Goldstein, 2001, p. 295).

There may be a dearth in empirical research, as Goldstein suggests, but some studies indicate a troubling connection between game-play and behaviour. For example, research conducted at the University of Aachen in Germany suggests that the brains of players of violent games react as though the violence were real and that, 'as violence became imminent, the cognitive parts of the brain became active and that during a fight, emotional parts of the brain were shut down' ('Brain Sees Violent Video Games as Real Life', 2005). While additional studies are required in order to gain a more conclusive picture of the possible effects of gaming on behaviour, the popularity of games suggests that violence and interactivity are a highly potent combination – a circumstance that has not escaped military strategists.

Playing war

Computer games have become a salient medium used to prepare individuals for combat. In some cases, the difference between a military version and a civilian game version is miniscule. For example, the game *Full Spectrum Warrior* (2004) has been published in two versions: one for the military and a slightly modified form for the public. The commercial version became a bestseller. Similarly, the game *Real War* (2001) exists for public entertainment as well as for military training purposes (the training version is called *Joint Forces Employment*). According to a review (Sieberg , 2001), 'the only difference between the two versions is that the official one contains more learning objectives and the player only has a finite number of military resources – tanks, planes and battleships. Visually, the game-play is nearly identical'. Also popular has been *America's Army* (2002), a first-person shooter game designed for recruiting purposes, which has reportedly been downloaded via the internet by more than 10 million people (Thompson, 2004b).

The similarities in gaming and military training products can help explain why, in Michael Moore's 2004 documentary on the Iraq war, *Fahrenheit 9/11*, an American soldier notes that he anticipated the war to be more like a computer game. Likewise, in the US bestseller *Generation Kill* (Wright, 2004), which documents American soldiers' experiences in the Iraq war, a Marine makes a reference to the controversial game *Grand Theft Auto*. 'I was just thinking one thing when we drove into that ambush ... *Grand Theft Auto: Vice City*. I felt like I was living it when I seen the flames coming out of windows, the blown-up car in the street, guys crawling around shooting at us. It was fucking cool' (unnamed soldier, quoted by Wright, 2004, p. 5).

In critical inquiries into gaming content, the question of narrative has served as a fruitful point of entry. Scholars have investigated the generally unorthodox storylines encountered in games and have compared them to traditional narratives, such as encountered in print or film. Others have challenged such comparisons; Celia Pearce (2004, p. 144) argues that, 'although there is much to be learned from traditional narratives, without understanding the fundamental difference, the discourse becomes ultimately irrelevant because it entirely misses the fundamental point of what games are about'. To shed light on 'the fundamental difference', and how we can interpret especially violent gaming narratives, a noteworthy *lingual* difference demands attention. At their core, games and the interactive narratives they feature are based on computer code, whereas traditional narratives are based on our natural languages. Pearce interprets gaming narratives in a 'play-centric' rather than a storytelling context. This approach makes sense as it directs attention to the computer languages that facilitate and enable gameplay. Traditional as well as new narrative forms inherit their structures from language. When language changes, narrative structures vibrate like a string that has been plucked.

Coding narratives

Whereas nonliterate cultures tend to view time in terms of organic rhythms, as cyclical as nature, in alphabetic cultures, despite the absence of scientific evidence, a linear model of time and linear narrative structures prevail. The pillars of narrative – beginnings, middles and endings – mirror the past, present and future tenses of language. Marshall McLuhan (1994, p. 85) notes that 'only alphabetic cultures have ever mastered connected lineal sequences as pervasive forms of psychic and social organizations'. We tend to read stories in a continuous manner; that is from beginning to end. The printed page – one letter placed next to another, lines of words consistently following in the same direction – reinforces a sequential and uni-directional narrative structure. Language, narrative structures and our conception of time are inextricably interconnected.

Computer games tend to consist of repetitive activities, such as making a game character jump, kick or run down endless corridors, across redundant terrain, round and round again at the racetrack. In the games *Doom* and *Quake* (1996), as in countless similar games, the repetitive action, aside from running, is shooting. There is hardly a moment of rest as the player is enticed to engage in the activity of pulling the trigger and aiming the gun barrel at anything that moves or may lurk in the shadows. Whatever the main theme of a game, this action will take place incessantly.

Repetitive events on the screen are the product of computer languages. Until a player decides to actively engage in game-play, the characters on the screen can often be seen standing at the ready, and appear to be breathing. Unless a player sends an instruction to move the character, the character will continue to stand 'ready' while behind the screen, computer languages send the instruction, 'stand ready and repeat the breath loop unless instructed otherwise'. Once a player decides to discharge a character's gun, a different part of the programming is read and an instruction is sent to execute the command, 'discharge gun'. For as long as a player makes the same choice, such as continuously pushing a button, the same lines of code will be executed and the same loop of action will occur on the screen.

Another way to understand how the story of a game is encapsulated in code is to consider that each action a gaming character can perform on the screen is contained within what in programming lingo is referred to as a routine, a piece of computer code that handles a task in a program. Penny (Wardrip-Fruin and Harrigan, 2004, p. 80) suggests that an analysis of interactive media must go beyond theories of representation and that, 'the content is as much in the routine that runs the image as it is in the image itself'. In other words, code defines the content a player sees and interacts with on the screen. It occupies the territory of narrative that was previously defined in literary or natural languages.

At this point it should be pointed out that linguists as well as computer scientists have put an emphasis on the differences between technology's artificial languages – code – and natural languages. Unfortunately, classified as artificial languages, computer languages have also been identified as 'un-human' languages (Ong, 1982, p. 7). Such categorizations can be misleading and are, moreover, contestable. Tom McArthur (1986, p. 69) writes: 'The structuring of books is anything but "natural" – indeed, it is thoroughly *un*natural and took all of 4000 years to bring about.' Calling language categorizations into question is not to be equated with disregard for the discrepancies between artificial and natural languages. Therein lies the crux; it is exactly because code constitutes a body of unusual languages that narrative in computer games presents the viewer with atypical story structures.

Where the elements of narrative – formal beginnings, middles and endings – once existed, repetitive action has taken over. Paul Virilio (Der Derian, 1998, p. 141) points out that in computer languages, past, present and future tenses have collapsed into a binary concept of events taking place 'now', or events taking place in the past. The structure and functionality of a computer game is sustained by programming that, once the game has started, checks for a user's input. Does the player engage in activity now? If so, the game ensues; otherwise, the process of checking for input will continue. In the binary world of computer languages, the three pillars of narrative have, in a sense, no place to stand.

Perhaps not surprisingly, narratives in computer games have drawn criticism for their insufficient complexity. Janet Murray (1997, p. 51) comments on the lack of story depth in computer games and notes that, even in very popular games, the plot is often so thin that it is difficult, if not impossible, to adapt the games into successful movies. Murray argues that the narrative deficiency of computer games is caused by the close attention that is usually attributed to the visuals and the related technical issues as opposed to the quality of the writing of the story (Murray, 1997, pp. 51–6). In a highly competitive gaming market, the quest for increasingly sophisticated visuals and game strategies can be all-consuming. But what undoubtedly resonate in the precarious state of narrative in computer games are the languages that are used to write interactive media.

Where there is no buildup or progress, and as the loops of repetitive action suggest, closure is elusive. The resistance to closure finds its ultimate expression in the meaninglessness of death that is so common in computer games. A tagline for the game *Manhunt* (2003) states, 'They just killed James Earl Cash. Now they want to kill him again'. On the computer screen, death – the decisive gesture of closure – has lost all finality. Walter Benjamin (1969, p. 94) writes that death is the sanction of everything that the storyteller can tell, but game characters often come equipped with multiple lives, or those who have expired are reborn with the next game. In computer games, the storyteller appears to have lost the final word – in the most literal sense of the phrase. Unlike traditional story forms, computer games unfold according to the instructions that are encapsulated in code, not words that, when combined artfully, can convey a deeper meaning.

Because computer languages are designed with the intent to sustain functionality, the very concept of closure is contrary to the purpose of code. Functionality is not supposed to be temporary but ongoing. Thus there is no final chapter, no last page to be turned, no concluding scene in computer games in the traditional sense. Characters run down darkened corridors endlessly, shooting their way from one level to the next, without ever really arriving – the killing tends to be ongoing. One of the consequences, arguably, is that whereas even the bloodiest of war films is likely to incorporate moments of tenderness or sentimentality, computer games seldom seem capable of pathos in a comparable way. In film, when tragedy strikes, the declining tempo of the storytelling, or a moment of silence, may effectively evoke a state of empathy in a viewer. Most violent games, on the other hand, leave little time to pause and reflect, or to experience tragedy. Rather, the violence in computer games is uncannily pure.

To a considerable degree, the experience of empathy depends on an event a viewer has no control over. Ken Perlin (2004, pp. 13–14) points out that, when watching a film or reading a novel, an audience's ability to make decisions is suspended. Rather, the emotions an audience experiences depend on, for example, the sequence of events and the timing of a storyteller. Also, according to Perlin (2004, p. 14), in a computer game, a player does not relinquish this aspect of control or agency. Instead, a player's agency remains with him or herself; it is primarily the player's actions that control the script and determine the sequence of events. It is worth considering that because, in a computer game, code and agency are interconnected, so are code and the potential scope for empathy.

Encapsulating violence

Where death is rendered meaningless, violence can be dispensed limitlessly. On the surface of things, it will have no lasting consequences. J. C. Herz (1997, p. 183) proposes the following: 'Violence was the obvious first choice for a game premise: put a target in front of the player and have him shoot in its general direction. At some point the target shoots back until one of them dies. It's self-explanatory, interactive and highly entertaining. It's also damned easy to program.'

Three-dimensional, virtual space breaks down into a grid of X, Y, and Z coordinates. If we begin by identifying a random point in virtual space, followed by a second point, connecting the two points creates a trajectory. Herz is right in noting that programming point-and-shoot scenarios is an obvious choice. While gunfights might be particularly easy to program, there conceivably is more to computer languages and programming that lends itself to the creation of violent gaming content.

Computer languages break down into individual sets of instructions – commands. Commands are exact orders that check for a user's input and send precise instructions as to how the machine is to respond. In regards to human commands, Elias Canetti (1984, p. 304) comments: 'The oldest command – and it is far older than man – is a death sentence, it compels the victim to flee. We should remember this when we come to discuss human commands.' It is disputable as to whether or not computer commands qualify as anything other than machine instructions. Canetti

(1984, p. 304), however, seeks to avert such distinctions and maintains that: 'Beneath *all* commands glints the harshness of the death sentence.' Interestingly, human and machine commands share key qualities. The human command finds its most frequent manifestation in the military – the cradle of computer commands is the Second World War. Canetti (1984, p. 304) further suggests that a command admits no contradiction; it should be neither discussed, nor explained, nor questioned, but has to be immediately understood. The same holds true for computer commands; code does not permit for interpretation. There is a kind of totalitarian logic at work behind the computer screen that resonates in the form and content of computer games. Programming does not allow for errors of any kind. So much as a misplaced symbol or letter can result in the failure of an entire program. Sherrie Turkle (2003, p. 509) observes: 'In a video game, the program has no tolerance for error, no margin for safety.' Once a program is functional, computer languages are translated into binary code. The on-off structure of binary code embraces a kind of unambiguous conception of all or nothing, life or death. Turkle (p. 509) echoes game players when she states: 'One false move and you're dead.'

Conclusion

Technology's underlying languages, or code, should be a factor in the critical inquiries into correlations between gaming and violent behaviour. Code should be a factor in such inquiries not despite the fact that computer languages vary significantly from our natural languages, but precisely because technology-based languages differ so greatly from natural languages and yet embody the power to facilitate game-play. Correlations between language production and violence have already been established. McLuhan (1994, p. 107) submits that the alphabet shattered the bonds of tribal man and remarks: 'The power of the printed word to create the homogenized social man grew steadily until our time, creating the paradox of the "mass mind" and the mass militarism of citizen armies.' Correspondingly, Leonard Shlain (1998, p. 377) writes that every society that acquired the alphabet shortly thereafter has become violently self-destructive. Empirical research that illuminates the interrelations between code, interactive games and behaviour is desperately needed at a time of war and the post-9/11 political climate when games are used for military recruiting and training purposes.

8
Impotence and Agency: Computer Games as a Post-9/11 Battlefield

Henry Lowood

Immediate reactions

Immediately after the terrorist attacks on New York City and Washington, DC took place on 11 September 2001 game publishers shelved or delayed projects with images, plotlines or game-actions reminiscent of the events. According to one observer, the attacks sent the industry 'into a frenzy' (ConsoleWire.com Staff, 2001). On 12 September, Electronic Arts (EA), the world's largest game publisher, suspended *Majestic*, an 'immersive game' that blurred boundaries between game and reality through pervasive, even intrusive use of the web, fax machines and telephones. Its plot included unexplained bombings, but after 9/11 frantic phone calls were as painful a memory as bomb threats; EA explained that 'someone who was waiting for a call from a family member or friend (involved in the attacks) [might] get a call from the game' (quoting Brown of Electronic Arts, 2001). Westwood Studios, a division of EA, delayed *Yuri's Revenge* (2001), the eagerly anticipated expansion of *Command & Conquer: Red Alert 2* (2000), to revise packaging art that showed screenshots such as a surprise invasion of New York City, though not to remove missions involving attacks on Washington, DC and the Pentagon. Activision, Konami, Ubi Soft, Microsoft and other publishers delayed or altered games, often to remove images of the World Trade Center. They hoped to avoid 'stirring emotions unnecessarily', as a Ubi Soft press release (Gallagher, 2001) put it. Yet, players resisted these intentions. Microsoft erased the Twin Towers from Manhattan's skyline in *Microsoft Flight Simulator 2002*, a 'last-minute panic alteration' according to one reviewer, but players created their own patch to put the removed Towers back into the default scenery (Dale, 2002).

Publishers did not make changes to these games because they simulated terrorist attacks; they were frightened by emotional resonances with essentially random game moments conceived and created long before the attacks. In *America's Secret War,* George Friedman (2004, p. 131) described the American public after 11 September in terms of 'emotional riptides', a 'state of shock', rage, feeling 'helpless and vulnerable', and a 'moral crisis' of 'deep uncertainty'. Game publishers responded to the emotional vulnerability. Microsoft apologized if 'that comment [by flight instructors early in the 2000 version of *Flight Simulator* that it "would be cool" if you crashed into the Empire State Building] causes anyone any pain. In retrospect, it's not appropriate at this point, and obviously it was not intended to hurt anyone's feelings' (quoting Pilla, 2001). Even when games were not about terrorism, publishers cut fragments with negative associations: story moments, bits of dialogue or images of buildings that had been destroyed. These quick fixes could not prevent subversive play, however, like flying into public buildings or modifying games to relive the events of 11 September.

In contrast to the game industry, web-based flash games and game mods by players and independent programmers cut much closer to emotional issues. Rather than avoiding memories and images of the attacks, they used these games to express their rage, 'get back' at terrorists, or grieve publicly. A post on the AllSpark.net forum (Mighty Quasar, 2001) on 11 September voiced the motivation behind such virtual agency as news sources reported a fourth passenger aircraft heading towards Washington, DC: 'DAMMIT. Why can't I have superpowers? Just one day, that's all I ask.' Like nearly everyone else, gamers felt helpless. Game designers might follow the lead of depression-era filmmakers by providing diverting entertainment. A spokesman for the Ziff-Davis Media Game Group (BBC News, 2001b) suggested that, '[n]ow more than ever people need escapism. Games provide a kind of all-engaging diversion that even movies cannot'. True, some players escaped into game-play, but many others let off steam in multiplayer game worlds, simulated revenge through virtual agency in wargames and modded shooters, or created simple games as commentaries on the events dominating their attention. One savvy *New York Times* writer noted that, even if game makers held back projects, 'players have other ideas'. Many players felt that game makers gave in to terrorists by postponing or changing games (Gallagher, 2001). So they created their own patches, skins, mods and flash games in response to the WTC and Pentagon attacks.

These projects began to appear right after 9/11. Hours after the WTC and Pentagon attacks, online communities used massively multiplayer roleplaying games as a medium for reacting publicly to the attacks. In *EverQuest* (1999) and *Asheron's Call* (1999) players spend hours at a time in their virtual, in-game lives; RL (real life) intruded when these games issued news alerts via in-game text or player community websites. Players organized candlelight vigils for the victims, using glowing weapons or other objects as candles. Screenshots documented a vigil held on the Luclin server on 12 September after 'yesterday's disheartening display of events'. According to one player, '[j]ust because you are in a game doesn't mean the world outside doesn't effect[sic] you. Many people would like to mourn and share peace along side people they have battled long and hard side by side with. Yes, I can go to a church to mourn, but I would like to do it with my comrades around the country/world, which is impossible everywhere else.' Thus, players were invited to 'mourn and discuss' on the Everlore website (Nirrian and Keeter, 2001).

Only days after the attacks, a group of players released a mission for *Rainbow Six: Rogue Spear* (1999) set in Afghanistan with the objective of hunting down Osama bin Laden (Gallagher, 2001). Osama skins were available just as quickly for most PC-based first-person action games. Players also created missions for realistic 'tactical shooters', such as 'Afghanistan' made for *Operation Flashpoint* (2001): 'Kill the terrorist leader ... Good Luck! It is not an easy mission – But not impossible!' (DANBAT_BillyTheKid, Axleonline, n.d.) Because mainstream commercial games require 18 to 24 months of intensive development, the game industry could only respond to 9/11 negatively, by delaying games or eliminating content. Players and independent programmers had other options. Mods like 'Afghanistan' granted virtual agency and superhero status through game-play. These projects could be completed quickly relative to building a new game and provided a perception of accuracy through representational verisimilitude. Most of the earliest 9/11 games were interactive cartoons created even more rapidly for dissemination on the web, often by using Macromedia's multimedia authoring program, Flash, available by Macromedia's estimate on 97 per cent of internet-enabled desktop computers. Websites such as Tom Fulp's Newgrounds. com ('The Problems of the Future, Today'), begun as a fanzine spinoff in 1995, distributed these games. By the end of 2004, this site consumed 500MB per second of bandwidth, most of it devoted to the distribution of free Flash games and movies (Fulp, Newgrounds (n.d.)). Newgrounds alone distributed perhaps 100 'anti-Osama games and movies' in the 'Osama bin Laden' collection; many more could be found on other websites.

If *Flashpoint* modders touted realism and detail, flash programmers were inspired by arcade simplicity. Amazingly, on 11 September there was already a game that let players defend the World Trade Center: *Trade-Center Defender* (also known as *WTC Defender*) at Lycos' hosting service, called Angelfire Arcade. Like *Space Invaders* (1978), *Missile Command* (1980) and of course Williams' *Defender* (1980), one only postponed the inevitable, shooting down as many aircraft as possible before some got through to destroy the Twin Towers. By 14 September, Lycos apologized for the game and pulled it from Angelfire Arcade for violating Angelfire's 'terms of service' (Angelfire Website, n.d.). A Bulgarian Internet Club was able to download the game before it was withdrawn, however. By 28 September, Stef and Phil's *New York Defender* emerged on a French-language site similar to Newgrounds; called Uzinagaz, this site possibly had distributed *WTC Defender*, which clearly inspired *New York Defender* (BBC News 2001a).[1] The new game used images of the burning Twin Towers etched in the minds of anyone who had seen them on television or streamed on the web. Jonathan Pitcher from Uzinagaz described it as a 'release' and a means for fighting 'our feeling of impotence. We reacted to September 11 like kindergarten children, by drawing planes crashing into buildings'. *New York Defender* appears to have been downloaded more than a million times from Uzganaz alone. Yet, rather than giving players 'a sense of excitement or joy, instead, it makes them feel powerless' (Thompson 2002; McClellan 2004). *New York Defender* projected impotence, not power.

A transition from anguish to rage can be observed in Flash games created after the 9/11 surprise attacks. Besides *New York Defender*, sites such as Artbeel.com, Newgrounds and Uzganaz collected 'Flash Fun Stuff', as Artbeel calls them, like *Bin Laden Liquors* and the *Kill Osama Game*. Rather than suppressing recollection, these games appropriated images such as Osama's face to engage in repetitive revenge fantasies. Hardly the stuff of military simulation or commercially acceptable entertainment, such game cartoons crossed turned despair and feelings of impotence into projections of unsparing revenge.

The October 2001 issue of *Computer Games* magazine probably reached newsstands and homes about two weeks after 11 September. Put to bed before the attacks, it included advertisements that had become eerily timely. A two-page spread for *Operation Flashpoint* proclaimed that 'This is war' and cited the feature of being able to 'create and share missions with the built-in Mission Editor'. Modders used this feature to create missions for hunting down terrorist leaders. The *Black Thorn* expansion of *Tom Clancy's Rainbow Six: Rogue Spear*, also advertised in this issue, illustrates the position of commercial anti-terrorist games in the weeks after 9/11. The original *Rainbow Six* (1998) established the tactical shooter and squad-based, anti-terrorist game as mainstream genres. The developer was Red Storm Entertainment, founded in 1996 by Tom Clancy and his partner, Virtus Corporation. Clancy had little to do with game design other than bestowing the 'Rainbow Six' title on the series, which has been a consistent commercial success (Upton, 1999; 2004). Red Storm promises 'complete immersion in the war of tomorrow' (Red Storm Website). *Rainbow Six* presented missions such as hostage rescues and anti-terrorist strikes, with a more focused emphasis on details that invoke a sense of realism than other first-person shooters such as *Counter-Strike*.

On 17 September 2001, the French parent company, Ubi Soft, announced it would delay *Black Thorn* 'out of respect for the tragic events that took place on Tuesday, September 11th'. The game must have been all but completed, but Ubi Soft would revise it 'to avoid stirring emotions unnecessarily and unwillingly offending the public'. The press release added that the game would offer 'a very strong moral component and realistic portrayal of the fight against terrorism' (Red Storm Website, news release). Completed on 17 October, it was in stores by the beginning of November. Ubi Soft broke with the reluctance of other commercial game publishers by locating the merits of *Rogue Spear* in its particular notion of a 'realistic portrayal'. It was not just *about* the war on terrorism, but took a stand *with* it. No game was better suited to the task. Brian Upton, then Red Storm's chief game designer, has revealed that the original story for *Rogue Spear* in 1998 featured a terrorist leader modelled closely on Osama bin Laden. The game's producer only altered the character in 'that the villain wasn't an Islamic fundamentalist' (Upton, 2004). Locations included a hostage rescue in New York City, a hijacked airliner and a rescue mission set in heavily bombed-out Djakovica, Kosovo, which one reviewer found 'pretty disturbing [...], since this is the stuff I used to see on the news programs – people killing each other in the silence of a wartorn town' (Mr. Domino, 2000). When players mobilized their elite team of international anti-terrorist operatives in *Rogue Spear*, in a sense they were already fighting bin Laden.

They could also modify *Rogue Spear* themselves. Just three days after the 9/11 attacks, an Australian mod designer with the handle 'Akira_Au' posted *Operation Just Reward* to his popular AtWar (@ war) website (Akira 2003).[2] Founded that April, AtWar was a gathering place for the 'tactical warfare gaming community', where players could download maps, mods, 3D models, skins and other resources to create new content for military games such as *Operation Flashpoint* and *Rogue Spear*. *Operation Just Reward* was put together by a distributed network of players working together over the internet; Akira conceived the mod on 11 September while 'watching the terrorist attacks', starting work on it as 'a way to vent the anger and frustration'. He described the mission as 'Delta Force is sent to Afghanistan to take down a leading terrorist'. *Rogue Spear* supported player-developed mods, but surprisingly perhaps, Red Storm and Ubi Soft were ambivalent about projects such as Akira's. Whereas Sierra, the publisher of Valve's *Half-Life* (1998), stated unequivocally that 'we definitely support the mod community', a spokeswoman for Ubi Soft stated, '[t]he company does not endorse or support the making of mods for its games' (cited by Kilgannon, 2002; see also: Wargamer Website, 2002). Andrew Baye, an American working with Akira on *Operation Just Reward*, defended the mission as providing players who 'feel powerless to act' with 'a chance to vent'. Ubi Soft distanced itself from the mod, however, and Akira removed it from his website on 20 September, citing 'strong negative emotions' awakened by the game (Gallagher 2001; Kilgannon 2002).

Despite the mixed acceptance from game publishers, mods such as *Afghanistan* and *Operation Just Reward* must have contributed to the surge in popularity of tactical shooters and related wargames immediately after 9/11. *Gamespot* reported on 27 September that *Operation Flashpoint* was the best-selling PC game for the week of 9–15 September on the US charts, a very high ranking for a tactical shooter, above such popular games as *The Sims* (2000), *Diablo II* (2000), and *Madden NFL 2002*. *Rogue Spear's* sales rose by 48 per cent in September 2001, according to *The New York Times*. The editor-in-chief of Game Monkeys, another game site, noted 'a massive increase in the desire to play anything antiterrorist, anti-evil-empire' (Kilgannon 2002). While the *Rogue Spear* advertisement in *Computer Games* proclaimed that 'terror has a new target [...] Team Rainbow', in fact it was Osama bin Laden (and soon, Saddam Hussein) who had walked into the sights of gamers. Games such as *Rogue Spear* and *Operation Flashpoint* encouraged players to vent emotions through the virtual agency of game-play. After 9/11, the player community used available games to create their own narratives, converting hostage rescues into assassination missions, for example. *New York Defender*, *Bin Laden Liquors* and *Operation Just Reward* momentarily addressed feelings of impotence and grief right after the attacks.

The entertainment industry, academic institutions and the US military cooperated in the development of training simulations and games about the war on terrorism, as well. The origins of this 'military-entertainment complex' go back to the late 1970s (Lowood and Lenoir, 2005). By late 2001, certainly during the second Gulf War, there could be little doubt that military games based on collaborations of commercial designers and the military were providing both entertainment and military applications. The military-entertainment complex provided players with virtual missions addressing new contexts of recruiting, training and battlefield execution that coincided with television spectacles of 'shock and awe' (a phrase Sony sought briefly to trademark on the day after US and allied forces marched into Iraq in April 2003) and embedded journalists reporting from the road to Baghdad on evening television.

Consider a small sampling of recent warfighting games: *America's Army, Rainbow Six: Rogue Spear, Full Spectrum Warrior* and *Close Combat: First to Fight. America's Army*, launched on 4 July 2002, was the official US Army game built for the Army's Office of Economic and Manpower Analysis; the team of mostly civilian programmers and artists at the Modeling, Simulation and Virtual Environments Institute (MOVES) used the commercially licensed Epic Games' *Unreal* engine to produce its graphics and game-play. *Rainbow Six: Rogue Spear* was a product of the multimedia entertainment industry. Shortly before 9/11, LB&B Associates, a military contractor in Maryland, obtained rights from Ubi Soft to modify the game on a Department of Defense contract for training American soldiers in urban warfare, focusing on small-unit decision-making skills in keeping with the game's design. The project director praised its accuracy: 'no game engine comes close to the realism of Tom Clancy's *Rogue Spear'* (Ubi Soft press release 2001). LB&B showed off its version at the 'Warfighting Readiness through Innovative Training Technology' (I/ITSEC) conference held in Orlando, Florida, in November 2002. By early 2003, the Institute for Creative Technologies had taken over the project and was using *Rogue Spear* 'in a super-powerful version' for an Army-funded project that the Army's chief scientist for simulations, Michael Macedonia, said would 'work brilliantly to sharpen decision-making skills at the small-unit level'. In October 2001, the ICT had begun developing *C-Force*, a squad-based game, and *Combat System XII*, a command simulator for company commanders (Kennedy, 2002; see also Trotter, 2003; Schachtman, 2001). These games became *Full Spectrum Command*, a PC-based training game that modelled Military Operations in Urban Terrain (MOUT) in Eastern Europe, delivered to the Army in early 2003, and *Full Spectrum Warrior*, developed with Pandemic Studios for the Army, but also published in commercial versions for consoles and PCs between March and September 2004. *Full Spectrum Warrior* depicted scenarios set in the Middle East. Destineer Studios and Atomic Games released *Close Combat: First to Fight*, based on military training tools, doctrine and input from the US Marine Corps. It continued the *Close Combat* series of military simulation games begun in 1996, spinning off from Atomic's *Close Combat: Marine*, developed for the Marine Corps. Ads in game magazines told prospective players that 'this is the tip of America's Military Spear. You're on Point'. Its selling point was that as one of the 'first to fight', *you* will be taken to the 'front lines of urban combat in Beirut', utilizing the Corps' 'ready-team-fire-assist' tactics 'now used in Iraq and Afghanistan' (First to Fight Website).

The confluence of commercial computer games, mobilization of the military-entertainment complex, the impact of 11 September, and the Afghanistan and Iraq campaigns has produced a deluge of both commercial and military games, connected to a Whiggish sense of inevitability about the capacity of military simulations such as 'Urban Resolve' to replicate post 9/11 battlefields in all their complexity. Mods and web-accessible games made for Flash or Shockwave players – the same sorts of game platforms that players used to hunt down bin Laden right after 11 September – have raised questions that military simulations had not tried to answer. These projects range from artist Josh On's *Antiwargame* (2003), a Futurefarmers project, to Fabulous 999's disturbing and nihilistic *Suicide Bomber Game* (originally, *Kaboom*, 2002). Rather than project virtual agency, these games undermine not just catharsis through simulation. Josh On (n.d.) noted that, '[t]here are multiplexes full of war propaganda like *Black Hawk Down*, and the Army has put out a recruitment war video game. I think this demands some symmetry'. The *Antiwargame* was supposed to counterweigh what On perceived as the political agenda behind military games.Yet, Fabulous 999 says only, '[b]y the way, I'm not Jewish, I'm not an arab, and I'm not a terrorist. I have little interest in what goes on in the middle east so I don't share any extreme views. I just think people who blow themselves up are stupid. That's all this game is' (Fabulous 999 – Newground, 2002).[3] The Uruguayan game theorist and designer Gonzalo Frasca expressed the limits of technology in the war on terrorism by simulating impotence, rather than agency, in his *September 12th: A Toy World* (2003). Rather than immersing players in a simulation of the urban battlefield, Frasca questions both the agency and the models behind such wargames and predicts the ultimate failure of the policies behind them. Its learning moment occurs through this failure of virtual agency; it is a simulation of failure. It is precisely this frustration of players that creates the message in *September 12th*, the *Antiwargame, New York Defender, or 9/11 Survivor* (2003), a mod based on a version of the *Unreal* game engine about escaping the burning World Trade Center (see 9/11 Survivor Website, 2003).

These projects reconfigure the computer game as a contested space. Clive Thompson has suggested that Flash and Shockwave games are merely online graffiti for the digital age, scrawled temporarily in cyberspace. But Frasca counters that, '[f]or political video games, September 11 was the trigger [...] If it had happened in the sixties, people would have grabbed their guitar and written a song about it. Now they're making games' (Thompson, 2004; McClellan 2004). Even when political goals are contested, the game as medium has not been. In 2002, the Syrian publisher Dar al-Fikr published *Under Ash*, followed in 2005 by a sequel, *Under Ash 2: Under Siege*. Afkar Media, the Damascus-based developer, asserted that, 'when you live in [the] middle-east you can't avoid being part of the image, as a development company we believe that we had to do our share of responsibility in telling the story behind this conflict and targeting youngsters who depend on video games and movies [...]' (Under Ash Website, 2004). Anyone who has played *America's Army* knows immediately how to frag Israeli soldiers in *Under Ash*. The developers of *Under Ash* did not resist the game form of *America's Army*, they appropriated it in order to depict 'military actions performed by local fighters against occupying forces'. For every *America's Army*, an *Under Ash*; and for every *Full Spectrum Warrior*, a *Special Force*, Hezbollah's shooter about 'freedom fighters with legitimate grievances' that the *New York Times* called '[t]he hottest video game for the teenagers of Beirut's southern Shiite neighborhoods' (Wakin, 2003).

In early 2005, Iraqis played New York-based Kuma Reality Games' *Kuma/War: The War on Terror*, an 18-mission campaign built with the cooperation of returned US Marines from the war in Iraq. They assembled in a café especially to play the mission that re-enacts an assault by insurgents on a Fallujah police station defended by poorly trained Iraqi police officers, 17 of whom died in the real assault. The café owner noted that 'the only people who don't approve are the resistance fighters themselves. One came in and told me that we shouldn't play games where we pretended to be US or Iraqi forces fighting them'. And yet the café owner's customers played on, as had the US Rangers who downloaded Akira's *Operation Just Reward* right before embarking for Afghanistan before them (Freeman, 2005; Werde, 2004). Military games published since 9/11 have taken on a burden beyond that of simulating events; they have become a medium for responding to an environment of threat and uncertainty.

Notes

1. The earliest surviving references to *New York Defender* appear in Belgian and Italian Usenet discussion boards in Google Groups, beginning 28 September 2001.
2. The *New York Times* reported that it was released on Sunday, five days after the attack.
3. By May 2005, this page had been viewed more than one million times.

9
S(t)imulating War: From Early Films to Military Games

Daphnée Rentfrow

Perhaps more than any other subject and object of representation, war has a tortured relationship to both mimetic and poetic production. Is the war story a narrative of events, casualties and experiences of battle? Or does the war story inform and shape the very act of war? Is 'realism' the most important criterion when evaluating a representation of war or do the stories we fabricate shape our experience and comprehension of war? Perhaps the connection between war and representation exists because, as Elaine Scarry (1985, p. 62) has argued, war itself 'has within it a large amount of the symbolic and is ultimately [...] based on a simple and startling blend of the real and the fictional.'

Perhaps the relationship between war and narrative exists because, as Clausewitz famously declared, war is not an isolated event but rather an expression:

> We deliberately use the phrase 'with the addition of the other means' because we also want to make it clear that war in itself does not suspend political intercourse or change it into something entirely different. [...] Is war not just another expression of [a people's] thoughts, another form of speech or writing? Its grammar, indeed, may be its own, but not its logic (1976, p. 604).

War does not merely happen to us. Rather, we *make* war in much the same way that we make literature and art, make love and make peace. We must pay close attention to the ways in which war (and the war story), in the words of Chris Hedges (2002), 'is a force that gives us meaning'.

But war is more than a semiotic element to be deconstructed or hypothesized – it has as its goal the injuring and destruction of human bodies. This is the one reality. But this reality is quite specifically

87

made – fabricated – in representations of war. Whether in literature, art, photography, film or computer games, death is staged in an attempt to render war's violence, to make us see and feel how war 'really is' though even the most sensory-laden reproduction can never duplicate war unless viewer-participants actually run the risk of being killed. Death is staged in visual representations because any other solution would be a type of snuff film in which the actors playing soldiers are actually killed on screen, and even then the death would not be a war death but a death meant to *take the place* of a war death. Virtual death in computer games is simply the most recent iteration in a long tradition of war representation. This essay, in turning to early film, maps how our new technologies of simulation aspire to a fidelity between the representation and the reality of war that is not so new as it is sophisticated, not so threatening as it is familiar, and not so successful as it is stimulating.

Analogue

If the collusion between war and industries of representation can be traced back centuries to the earliest forms of writing and art, it is also true that the 19th century in particular witnessed a happy marriage between modes of creative production and militaristic destruction. Photography emerged as the most capable of art forms, capturing life 'as it happened', including the life of war; chemists experimenting with the nitrocelluloids of explosives discovered that the new emulsions could be used to fix images to film; the hand-cranked machine gun inspired the moving picture camera; the Boer War, coming only three years after England's first display of the cinematograph, helped launch the 'Boerograph' and the 'Wargraph'. Photography and film, those most modern of inventions, shared the stage with that other most modern character, the battlefield.

Photography and war shared a symbiotic relationship from the former's first appearance.[1] In 1839, a commentator on the newly invented photographic process of Louis Daguerre celebrated the new medium's ability to render landscapes: '[…] as three or four minutes are sufficient for execution, a field of battle, with its successive phases, can be drawn with a degree of perfection that could be obtained by no other means'.[2] Photography was as well-suited for the documentation of war as war was for the promotion of photography's own technological wizardry. Daguerreotypes and calotypes survive from battlefields as early as 1846, and since then war photography has become both journalistic partner and lucrative public spectacle.

Yet early photographic technology did not allow for 'action shots'. Pictures could not be fixed in fewer than 20 seconds, so the only images that were captured were by necessity static; in times of war this trans-lated to images of soldiers posing together, war-ravaged landscapes, the wounded and, of course, the dead. Mathew Brady and his associates cap-italized on the public's desire for war images while mastering the static pose. Brady's exhibit 'The Dead of Antietam', with its images of disfigured corpses and wasted landscapes, was a great success. The photographs (taken largely by Alexander Gardner) brought home the images of modern war with a new terrible immediacy and earnestness. Yet while the images from Antietam were of the dead where they lay, Gardner and others in Gettysburg ten months later moved bodies around to achieve the par-ticular shots desired: tableaux were arranged, and darkroom manipulations rearranged the scenes even further (Marwil, 2000, p. 34). The documen-tary imperative that necessitated actual footage dovetailed with the artis-tic and financial imperatives to commodify and sell that same footage.

In early war films these staging techniques were transformed into live action re-enactments. Diffuse fields of engagement, blinding dust and smoke, mass armies grouped across wide fields of vision, and the speed of industrialized death proved too much for heavy and cumbersome cine-matographic machinery. In order to represent modern war, then, mod-ern filmmakers had to resort to re-enactments, fakery and staging.[3] Toy ships in bathtubs, battles shot against painted backdrops and orches-trated infantry charges became the *sine qua non* of these early films which were, ironically, marked as modern by their *very ability* to (convincingly) capture battle. The spectator of staged war films was immersed in fan-tasies of visual and physical dominance while learning what modern war *looked like*. That what war looked like was in fact *fake* meant as little to the viewer as did the rearrangements of corpses on the battlefield in war photography. The *reality effect* of the war film was to transform re-enactment into 'the real thing'.

War films are still evaluated in terms of their realism, and the spectator is taught what war looks like *while all the while* marvelling at the sophis-ticated fakery that makes the war look real. These early re-enactments prepared the spectator for later narrative films in which location shoot-ing amplified the realism of battle scenes, even if the location was a California beach masquerading as a Vietnamese jungle. The outbreak of the First World War, which would coincide with a turn to narrative film, continued and formalized the trend of the war film but introduced one significant new detail: live-action death.

In 1916, the Committee on War Films sent two cameramen to the front to get footage of the war; these men eventually captured the advance at the Somme. Originally intended as a newsreel, the footage was edited and censored and finally made into a full-length 'documentary' entitled *The Battle of the Somme* that ran 75 minutes, with text supplied by the War Office. It opened on 21 August in 34 cinemas in London and by early September more than 1,000 theatres across England had booked showings. Upon viewing the film, the King declared that 'the public should see these pictures that they may have some idea of what the Army is doing, and what war means' (cited by Hynes, 1991, p. 123). The film offered the war as celluloid object-lesson. It represented what the Army was *doing* while defining what the war *meant* and what *it looked like*.

Yet, as was the case with the Boer War and the Spanish-American War, the First World War offered little in terms of cinematic action. Hampered by heavy cameras, limited lens depth of field and censorship, the cinematographers at the front produced little other than, literally, moving pictures. With limited equipment (a stationary hand-cranked camera on a tripod) and limited possibilities (the cameras could not zoom in or out), exacerbated by government restrictions on what could actually be filmed (dead British bodies being disallowed), the filmmakers could not come close to capturing an attack of 13 divisions across a 16-mile front. The camera's immobility prevented it from going 'over the top' with the soldiers and the violence of No Man's Land made recognizable shots of the enemy impossible. The results were medium-centred shots of companies marching to the front, men in the trenches moving about before battle, and the wounded returning. In *The Battle of the Somme* there is no narrative line, no central characters to follow, no love story to counter the (in)action of battle. And yet the film was celebrated and remembered for its riveting action.

This apparent paradox is understood once it is clear that only one particular moment is the best remembered from the film. It is that moment that makes the footage *real* in a way that was new to film; it is the moment when the viewer sees 'what war means' to a British soldier – death. The scene in question is fairly unremarkable to a modern viewer: soldiers scramble out of a trench; two fall back, one slides down the side of the parapet; the scene shifts; a low angle shot reveals more than a dozen men advancing through barbed wire; two more men fall. This, the moment of death, is the most memorable scene, the climax in a film without climax. It is also the only scene in the film that was staged.[4] The viewers had not *seen* death in battle; they had seen a *representation* of death in battle. Yet it is this scene which was to become the most memorable of the film.

The audience, trained by two years of war and personal experiences to know that death was neither quick nor graceful – men had already returned to England with limbs missing, devastating psychological damage, disfiguring facial wounds and stories of bodies literally blown to bits – still responded to the scene as if it documented something real. The auratic quality of the filmed image suspended disbelief, rendering the scenes 'accurate' and even 'gruesome'. More importantly, however, is that *The Battle of the Somme* differed from its contextual peers. It is a documentary produced when cinema was eagerly moving away from the cinema of *actualité* in favour of organized narrative. *The Battle of the Somme*, then, was clearly differentiated from fiction film. The movie essentially *looked* different. Rather than suffering from its similarity to 'primitive' cinema, *The Battle of the Somme* capitalized on its resemblance to an animated newspaper and became documentary at the moment that it became fiction. The re-enactment of death as the moment of *war's* climax made the film 'more real' than it would have been otherwise. The documentary imperative that necessitated a death in battle paradoxically provided *narrative* closure; the 'death' of the unknown Tommy became *the* story of the film.

The documentary imperative in war films is alive and well in contemporary cinema. Perhaps the best example of this is Steven Spielberg's *Saving Private Ryan* (1998), celebrated for its realism, its graphic nature and its scenes of the audacious assault at the beginning of the film. The realism of *Saving Private Ryan* is quite specifically a production on the grand scale, a production that, like all others, masks poesis as mimesis. There is no need for battlefield footage when the apparatus itself can produce more realistic images. *Saving Private Ryan* looks real precisely because it doesn't look like or use black-and-white footage from the Second World War, which in the aftermath of colour, improved sound editing and televised war suffers from its datedness. Suffers, in fact, from its own historicity.

Where once static images evoked the dynamism and spectacle of modern war, now elaborately choreographed battles supplant actual footage. Yet the documentary imperative still demands that the war film *look* 'real', and the most mimetically resonant images are still those that are fabricated. The biograph in battle has been replaced by computers and advanced cinematography but the subject still requires re-enactments to make the chaos of modern warfare visible, and the essential re-enactment is that of death.

Digital

If *The Battle of the Somme* and *Saving Private Ryan* are two extremes of the same tradition then computer war games are simply a new iteration of that tradition. The irony is that while films evolved to deploy more and more realistic renderings of battle and while computer games are routinely criticized for their violence, digital war in one particular arena has been moving toward cleaner and less violent depictions. When it comes to the United States Army, mimesis has become all about teamwork. Death, apparently, is not in the Army's interest.

The military's awareness of commercial gaming arguably began with the introduction of *Mech War* in the late 1970s to the Army War College, and has continued at an accelerated pace since the Department of Defense (DOD) recognized the strategic capability of simulation technology: wargaming and simulation training, for example, are now a part of the curriculum of every US war college (Macedonia, 2002, p. 6). Gaming technology allows the military to create sophisticated training modules while taking advantage of the media that young soldiers have grown up using. While arcade games were used in the early 1980s as 'skill-enhancers' it wasn't until 1996, when the Marine Corps Modeling and Simulation Management Office adapted the commercial game *Doom II: Hell on Earth* (1994) into *Marine Doom* specifically for training, that gaming in the military gained acceptance. The AI monsters in that game were transformed into opposing forces, and the game became about concepts such as 'mutual fire team support, protection of the automatic rifleman, proper sequencing of an attack, ammunition discipline and succession of command' (Macedonia, 2002, p. 7). Military interest in the possibilities of gaming-as-training increased as game companies such as Novalogic and Ubi Soft Entertainment began adapting their commercial games to help train soldiers. By 1999 these efforts had come together in guided research efforts to systematically explore the possibilities offered by the commercial entertainment industry for military training and education. In that year, under the direction of Michael Andrews, the Deputy Assistant Secretary for Research and Technology, the Army established the Institute for Creative Technologies (ICT) at the University of Southern California (Macedonia, 2002, p. 7).

Described on its website as a 'partnership among the entertainment industry, Army and academia with the goal of creating synthetic experiences so compelling that participants react as if they are real,' the ICT's research covers simulation and training, artificial intelligence, post-traumatic stress disorder assessment and treatment, leadership training for the Army, combat-control training modules, and much more. As the Hollywood screenwriter and ICT participant David Ayer (ICT website, 2004) describes one of the projects: 'You can create veterans who've never seen combat.' Since ICT brings Hollywood together with the Pentagon, it is not surprising that Michael Macedonia, chief technology officer for the Army's Program Executive Office for Simulation, Training and Instrumentation, describes the genesis of the $45 million merger in one question: 'The basic idea was, "Why can't the Army be more like Disney?".' The collusion between the military and the entertainment industry (what James Der Derian (2001) calls the 'military-industrial-media-entertainment network'), as troubling as it may be, merely renders visible what has for so long been the relationship between war and the imagining industries.

Nineteenth-century photography and early cinema relied on fabrication, staging and manipulation to represent war; even though live-action death was simulated, it at least aimed to represent some approximation of war's consequence. Increasingly, in the 21st century military, death is the *one thing* that is becoming beyond representation; nowhere is this clearer than in the popular *America's Army* (2002). This online game produced and promoted by the United States Army lets gamers 'play soldier' in detailed scenarios that the Army proudly claims are the most realistic of any battle-oriented games. Available for free downloading at americasarmy.com, the game boasts (according to its own site statistics), 'over 5 million players, over a billion missions played', and is one of the top PC online action games on the web. Colonel Casey Wardynski, the creator of *America's Army* and director of the Army's Office of Economic and Manpower Analysis at West Point, explains that the game, in which the Pentagon has invested $16 million, is more effective at delivering the Army's message to young people than the hundreds of millions of dollars the Army spends yearly on advertising (cited by Schiesel, 2005). According to Wardynski, the game is not explicitly used for recruitment ('the Army will not be able to identify you individually unless you choose to reveal your personal information') but 'players who request information [about the Army] *and* reveal their nom-de-guerre to Recruiters', however, 'may have their gaming records matched to their real-world identities for the purpose of facilitating career placement within the Army' (*America's Army* FAQ, n.d.). In other words, though not specifically a game designed for recruitment, it does have built into it that particular use:

> Data collected within the game such as which roles and missions players spent the most time playing could be used to highlight Army career fields that map into these interest areas so as to provide the best possible match between the attributes and interests of potential Soldiers and the attributes of career fields and training opportunities (*America's Army* FAQ, n.d.).

Yet while the game aims for realism in every detail, there is one area which is very specifically *not* realistic: at a time when computer games are increasingly criticized for their violence, *America's Army* limits the very thing that defines warfare:

> We built the game to provide entertainment and information without resorting to graphic violence and gore. When a Soldier is killed, that Soldier simply falls to the ground and is no longer part of the ongoing mission. And in the MILES laser tag mission, when a Soldier is hit, he just sits down and there is no puff of blood. [...] The game does not include any dismemberment or disfigurement (*America's Army* FAQ, n.d.).

As Chris Chambers, a retired Army major who is now the project's deputy director, describes the choice:

> We don't use blood and gore and violence to entertain. That's not the purpose of our game. But there is a death animation, there is a consequence to pulling the trigger, and we're not sugarcoating that aspect in any way. We want to reach young people to show them what the Army does, and we're obviously proud of that. We can't reach them if we are over the top with violence and other aspects of war that might not be appropriate [for the Teen rating]. It's a choice we made to be able to reach the audience that we want (cited by Schiesel, 2005).

This is the logic of *America's Army*, the same logic by which ICT applications can produce veterans who have never seen combat. War is not a game, we are told, and yet games are designed to produce soldiers ready for war in which death doesn't happen and blood and gore do not exist. *America's Army* acknowledges the impossibility of representing war by reducing death to a non-representation, a lack whose supplement merely creates another lack in an endless loop.

It would be tempting to end here with a criticism of the military's use of gaming, simulations and deathless battles as a Disney Factor that reduces reality to spectacle; it would be easy to describe *America's Army* and other examples of military gaming strategies as programmatic renderings of theories of the virtual. But when taken as a part of the history of war representation laid out here, the logic and consequence of *America's Army* is not so simply theorized. Whereas films have staged death in war in an attempt to makes themselves realistic, as if film were actually rendering the real of war, *America's Army* refuses this same representation in order to highlight the game's status *as a game*.

Der Derian, in a work similarly interested with virtual war, advocates a 'virtual theory' which 'seeks to understand how new technologies create the effects of reality [when] that reality has always been inflected by the virtual' (Der Derian, 2001, p. 217). In a strange twist of fate, this is where the logic of *America's Army* seems to me to reside: by refusing to represent death the game insists on its status as non-entertainment. While the gaming industry defends the violence in its games by insisting that the games and their depictions of violence are intentionally hyperbolic entertainment, the Army refuses violence because the games are intended to be about teamwork, tactical training and immersive experiences to help young soldiers learn the strategies of war. Simulated death has no place in these games because, unlike commercial games, they are not meant for entertainment – they are meant, quite specifically, as education. By refusing the gruesome, bloody, scream-filled moments typical of violent computer games, *America's Army* refuses to falsify death in an attempt to make the game realistic. Its bizarre realism, then, rests specifically in a refusal of realistic representation. Yet it is this (non)realism that is used to recruit new, often naïve soldiers. Jean Baudrillard's hyperbolic claim that the Gulf War did not take place seems more prescient than ever when we consider that a generation of American soldiers, raised on the violence of computer games yet recruited and trained in death-less Army games, are currently fighting the second Gulf War which, unlike its predecessor, is distressing to the American public precisely because the messy footage of it looks nothing like the world of games.

Death is the event horizon of war representations and without it everything else is simulation. For Der Derian, this is the most troubling aspect of increasingly 'realistic' simulations of war in both the gaming industry and the military use of simulations for training: 'In a sense, then, war has always been a virtual reality, too traumatic for immediate comprehension. But now there is an added danger, a further barrier to understanding it. [...] In this high-tech rehearsal for war, one learns how to kill but not to take responsibility for it, one experiences 'death' but not the tragic consequences of it' (Der Derian, 2001, p. 10). Der Derian describes this as a new danger particular to the advanced technological wizardry used in contemporary military training, in which soldiers perform manoeuvres in a hybrid world of real armour, location and matériel. This question of simulated death, however, belongs to a much longer tradition of fakery and re-enactments. Sanitized death simulation is as much a part of the training of modern armies as representing war is a part of our literature and art; increased technological sophistication simply changes the way death looks. And in that simulation, from Brady's Civil War photographs to ICT's wondrous applications, we find an attraction that, in the end, not only simulates war but stimulates it.

Notes

1. For details see Jonathan Marwil (2000).
2. Joseph Louis Guy-Lassac reporting to the French Chamber of peers, cited in Marwil (2000, p. 30).
3. Other scholars exploring the topic include Kristen Whissel (2002), who traces battle re-enactments from Buffalo Bill's Wild West shows to films of the Spanish-American War; James Castonguay (n.d.), who has mapped the various imperialist and racist ideologies motivating early filmic representations of the Spanish-American War in United States media culture, as has Charles Musser (1990); and Nicholas Daly (n.d.), who turns attention to the Boer War in early British cinema.
4. There is some debate as to whether or not other scenes were faked. Clearly, some scenes were designed that would allow the soldiers to acknowledge the camera as they walked cheerfully by. Others are suspicious. But it is widely agreed that the 'going over the top' scene is a re-enactment. For a shot-by-shot analysis and assessment of fakery, see Robert Smither's ' "A Wonderful Idea of the Fighting": The Question of Fakes in *The Battle of the Somme*' (1993).

Part IV
Ethics and Morality

10

Player in Fabula: Ethics of Interaction as Semiotic Negotiation Between Authorship and Readership

Massimo Maietti

The ethics of computer gaming are often approached as an issue of mere content, that is, a statistical account of the events that happen within the game world, and their analysis from an ethical standpoint (see Provenzo, 1991; Tuchscherer, 1988). Via the application of a fairly basic content analysis grid, the rates of violence, racial or sexual discrimination, and obscenity are assessed. The outcome is predictable: computer games tend to under-represent minorities, and to over-represent stereotypes and violence. But is this a valid methodology to asses the ethics of interactive media? Events, in computer games, take place through the presence of the user, whose role is not to merely actualize a single predetermined narrative, but rather to act within certain boundaries of performative freedom. The analyst's objective gaze, external to the fictional universe, might not be able to account for ethics in interactive media, because interaction unfolds only when the user enters the narration, acts in it and reacts to it. In other words, the specificity of the medium – that is, interaction – does not relate primarily to the events narrated, be them the rate or gruesomeness of killings, but rather to the performance that causes these events: the shooting itself, the act of shooting or the decision to shoot. The trigger is pulled by the player, and therefore the responsibility of the narrated acts is no longer a function of the relation between a questionably moral teller and a voyeuristic spectator/reader, because the simple but yet voluntary and self-aware gesture of pressing the button throws the player into a position of responsibility which is incommensurably different from the notions of ethical readership put forward in the field of ethical criticism. Schwarz (2001, p. 10), for instance, argues that '[r]eading complements our experience by enabling us to live lives beyond those we live and to experience emotions which are not ours; it heightens our perspicacity by enabling us to watch *figures* [...] who are not ourselves, but like ourselves'. It could be argued that the figures of interactive media might not be necessarily *like* ourselves – as games such as *Frogger* (1981), *Tetris* (1986), *Mercury* (2005) or *Katamari Damacy* (2004) show – but to some extent they *are* ourselves, for the decisive reason that players go beyond the screen and become part of the narrated universe as an entity of that universe. However, this should not lead to another, symmetrical, fallacy: to shift the whole problem from the narrator onto the user. In fact, to claim that the ethics of social responsibility have to be directly applied to player behaviour in fictional universes is another theoretical pitfall. Such an approach is based on the idea that real and interactive fictional worlds should share the same ethics (as it is implied, among others, by Adams and Rollings, 2003, p. 79; or Thompson, 2005), and therefore that the choices and actions of the player in the computer game should be directly compared to their hypothetical outcome in the real world. This assumption is questionable for at least two reasons.

First, computer games create fictional universes that, as such, can institute – and function according to – an ethical system different from that of the real world. Of course, this system can represent a commentary, a critique or a parody of current ethics in the real world, but the relation between the two ethical systems is not a simple equivalence.

The second reason is related to the role of the ethical subject. In computer games, the user, the one who is thrown into a fictional world, does not fully retain her or his individual properties and identity. In his classic study on the anthropology of play, Huizinga defines the game's *magic circle* as a 'temporary world within the ordinary world, dedicated to the performance of an act apart' (1950, p. 10). In the magic circle, the *homo ludens* takes on a new role according to the rules and the ethos of the game. Computer games are not merely tools for communication, in which the user is granted an unfettered liberty to use the specific language of the game/tool to create narratives, to conceive the rules that build the narrative worlds s/he inhabits, and to self-define her or his own identity. Unlike chat rooms or other communication systems, in order to generate a fictional identity in computer games and narrative-based interactive media it does not suffice to state the new identity ('Hi, I am a 25-year old single man') through the specific language of the medium. In narrative interactive media, the identity is not fully created, but it is at least partially inherited. The user cannot carry over her or his identity, because to become part of narrative textuality means to be born again in the fictional world, and to be confronted with its ethics and politics.

As Gee (2003, p. 44) argues, 'a given identity associated with a given semiotic domain relates poorly (or well) – in terms of one's vision of ethics, morality, or a valued life – with one's other identities associated with other semiotic domains [...] In this sense, then, semiotic domains are inherently political'.

Therefore the player inhabits the fictional world in the form of a third unit, different from both the user in the real world and the hollow character in the motionless pre-interaction fictional universe. This unit, the *simulacrum*, is the *bridge* between the two worlds as well as the subject upon whom the ethics of interactive textuality are founded: outside of the computer game, the user responds to the ethics of the real world; likewise, the fictional universe prior to interaction is a motionless landscape in which no ethical choice can be made.

Ethics is a discourse whose objects are the beliefs and behaviours of an individual or a group within a community or organization. In computer games, the very first ethical problem arises when this new, third, identity is formed. It is in this very process that the player's cultural ethics are confronted with the ethics of the possible world through the space of their overlapping, that is to say, the fragmentary unity of the *simulacrum*.

The first terms of this ethical negotiation are the users and their identities. Are the users of an interactive system *readers* or *authors*? Are they a *bricolage* of both figures? Or none of them? And what are their competences? And how are ethical issues related to users' roles and competences?

The issues of readership and authorship, as defined in literary theory (see Eco, 1979; Fish, 1982), can be discussed from the point of view of the ethics of reading and the ethics of the narrated events, respectively. Clearly, however, the user of a narrative interactive system is not a reader, as the events that occur within any game world are by definition partially caused by the presence, within that fictional universe, of the user's *simulacrum*. Symmetrically, the ethics of authorship, that refer to the subject of narration, cannot be directly applied to the user, for the mere reason that s/he is thrown into a possible world whose functioning rules are to some extent pre-assigned.

The authorship of an interactive system is in fact no less problematic than that of its readership: not even the entity (be it the actual game designer or development team, or the analytical figure of the narrator) that actually enforces the rules upon which the game is based can be considered as possessing the full set of competences generally assigned to authorship in literary theory, because those specific rules are not valid until a user deals with them, validating or opposing them.

Once again, in order to pursue the intellectual challenge of studying how the ethics are set and shared in an interactive system, player and game cannot be analysed as two ultimately detached dimensions.

In order to investigate the negotiation between the position of readership and that of authorship it is useful to employ a semiotic methodology, with a particular emphasis on interpretative semiotics and narrative theory of possible worlds.

Semiotics, as well as other disciplines that study how meaning is generated and communicated, is based on a sharp distinction between narrator and narratee, sender and receiver, author and reader. However, as I have tried to show, in interactive media these distinctions have to be questioned. Let us propose a *Gedankenexperiment*, in which A, a user, is playing, for instance, the horror game *Doom* (1993) on a computer, and the audiovisual output of the game is projected and diffused into another room, where sits B, a generic spectator. B will be staring at an audiovisual flow, and s/he will be in the ordinary position of a film spectator: no feedback on the visuals and sounds, a pure position of spectatorship. The question here is: if B is the empirical spectator of the *Doom* footage, who is its empirical author? B is spectating one of the countless possible *Doom* configurations, and this singular and specific sequential audiovisual text is an output of an interaction. This output exists by virtue of a specific property of interactive textuality: every fruition of a non-sequential matrix of narratives generates a unique, sequential, non-interactive text. While the flow of an online chat represents, for its participants, a regulated but open interactive communication system, its deposit, the chat log, is a piece of written dialogue; the navigation through the results of an internet search is a collection of documents that cover a common topic; the computer game *Doom*, once played, becomes a low-definition horror-thriller CG movie. However, this text, produced through the fruition of an interactive system, can be defined as *terminal*: it is the final outcome of a process of interaction. Likewise, B is the *terminal spectator* who is being fed with the audiovisual stream of the *terminal text*, an inevitable as well as accidental product of interaction. The fallacy of content analysis and other disciplines and methodologies consists precisely in analysing the *terminal text* as if it were an actual portion of interactive textuality. It is not. The whole process of interaction is designed around the intimate experience of the contact between the player and the fictional world in its becoming, in the openness of the different possibilities. When the interactive text is considered as a given object – a being no more in its becoming – its inherent aesthetic, narrative and also ethical goals do not belong any more to the domain of interactivity.

Interactivity, in other words, has to be analysed in its very unfolding, and not *ex post*. This effort reveals an epistemological fallacy already highlighted by postmodern thought: traditional sciences of meaning are based on the epistemological model of a text as an unmovable, subdued object, and cannot make sense of a polymorphic, undetermined, living textual entity created by a similarly undetermined author.

During the 1990s, two fields of study tried to tackle this very issue. In the classic debate on hypertexts, George Landow (1992, 1994, 1997) backed the idea of a *reconfigured author*, a weakened figure forced to hand over to the reader some degree of responsibility over the text. Once again, however, the object under scrutiny for such claims was the *terminal* text, the final outcome, and not the interactive narration in its becoming.

In the same period, Pierre Levy (2001) talked about interpreters, collective authors and works without author, and he ultimately challenged the notion of authorship. However, his conclusions, despite their relevance for the sociology of culture, are once again based on the epistemology of fixed objects created by some identifiable empirical author.

In order to proceed beyond the simple assumption that empirical authorship in *terminal* texts is a joined contribution of both the user and the designer of the interactive system, and to analyze more deeply this complex process in which events – and ethics that govern them – are generated in interactive fictional worlds, it is necessary to move from an enquiry on empirical authorship to a theory of Model Figures (Model User and Model Designer), in the same fashion as semiotics of narration – and the notions of Model Reader and Model Author – were developed in the field of literary theory during the 1970s (see Eco, 1979; Fish, 1982).

Interaction, defined as a process in its becoming, can be regarded as a dialectic between the potential, considered as the class of the possible configurations of the interactive system, and the actual, the specific sequence of configurations on a timeline – a dichotomy that closely refers to that of syntagm and paradigm in Hjelmslev (1961). In this process, the Model Designer sets the boundaries of the potential, whereas the Model User generates the actual, the evenemential.

It should be noted, however, that in narrative interactive systems, the realm of potential is not one of limitless freedom in which every combination is valid. On the contrary, the reduction from the potential to the actual (that is, the process of interaction) is defined by its constraints more than by its openness: in an example by Suits, the player who has to go out from a labyrinth will not tear its walls apart, even if they were paper-thin, but s/he will respect the boundaries they constitute, because the player's purpose 'is not just to *be* outside [...], but to *get* out of the labyrinth, so to speak, labyrinthically' (1978, p. 32). The pleasure derived from using the interactive system is not the pleasure of authorship, of creating texts out of a language, selecting and recombining tokens, but is the pleasure of acting and reacting to a realm of potential that can widen or narrow at each step. The potential is not a landscape of different configurations, but a systemic matrix of textual strategies that, when actualised, create effects of meaning, for instance reducing or magnifying the user's range of possible interaction at a given point. The pleasure of the user does not derive from the fruition of the actual, but from the fruition of the potential or, better still, from the transition between the two.

This transition can only take place on one condition: the user should possess some competences prior to his or her performance. Just as the Model Reader of a novel should be proficient in the language in which the novel is written and should have an encyclopaedia of notions and a specific knowledge of the literary genre (see Eco, 1979), the Model User of an interactive text should possess at least four kinds of competences:

1. Competences regarding the interface: the Model User should be able to interact with the system by means of perception (to receive the system's outputs) and expression (to provide the system with inputs). Secondly, the Model User should possess the ability to relate perception and expression to the change in configuration of the fictional world, so as to derive some rules for interaction, such as: 'if I press the red button, my alter ego walks'.
2. Competences in the real and the fictional world: the Model User should be able to conceptualise the fictional universe in its differences with respect to the real world (see Doležel, 1998; Pavel, 1989) thus inferring the mechanics that govern the possible world.
3. Axiological competences: competences regarding the mechanics of the possible world are based on the realm of cognition. Even if games are often analysed exclusively for their cognitive component (for instance, as a sequence of hypothesis and deductions – see Colombo and Eugeni, 1996), pathemic elements and values – in generative semiotic terms – play a very relevant role in interactivity, for they inform the nature of the player's involvement within the events of the possible world. Narrativity in interactive systems is not a storyline (do *Tetris* and *The Legible City* [1989] have a storyline?) but it is rather constituted by the values (and their exchange and mutation) at stake in the reaction of the fictional world to the actions of the user, in ethical terms. Good and bad, positive and negative, desirable and undesirable, increase and decrease: the user's actions are evaluated by the system

(sometimes with the most extreme form of reaction, the *Game Over*: the user is expelled from the fictional world) and s/he, in turn, has the faculty to align with the ethical system that emerges from this series of rewards/punishments, or to challenge it.
4. Competences regarding the role. When engaged in a computer game, the user, from the spatio-temporal position in the real world, enters another time and another space. This narrative *débrayage* (Greimas, 1983; Greimas and Courtés, 1982) is coupled with the shifting of identity: the user takes on a new identity as well.

This fourth kind of competence is of particular interest for the ethics of interactive media. In fact, it does not suffice to study the ethics and axiological systems of the possible world; it is necessary to analyse the role or the player's *simulacrum* as well. The role assigned to the user is in fact the result of a very complex and sophisticated semiotic process that establishes the primary dimension of *distance* between user and *simulacrum*. The fictional alter ego can resemble the player and convey a strong degree of isomorphism (short distance between the user and *simulacrum*), or it can drift away and not be commensurable to the user (long distance). The shorter this distance, the more the user is – and feels – ethically responsible for the actions of his or her *simulacrum*.

On the figurative layer of the interactive text, this distance is reduced if the *simulacrum* is represented as a human being, and even more so if the user can customise the physical and facial appearance of her or his alter ego, setting gender and race, adjusting the facial features and so forth.

Another level on which distance is negotiated is the point of view. In cinema as well as in computer games, the point of view of the camera is the point of view of the narrator. However, in computer games, the position of the camera also plays a role in the relation between the player and the *simulacrum*. Subjective or first-person camera usually prevents the user from seeing the avatar: this means that the *simulacrum* is a faceless, empty vessel. Their distance is at a minimum: they both look through the same eyes, and therefore what they see (and what they know) of the fictional world overlaps. The result of the subjective point of view, on the ethical level, is that the user will not be *playing as* someone else, because its mask, the alter ego, is see-through: the user will be directly responsible for the actions carried out in the fictional world. There will be a *transparency* of identity. On the other hand, an objective camera will create a wide distance between user and *simulacrum*. The user cannot see through the eyes of the alter ego, but s/he does see the alter ego as it is, in all its peculiarities and individual features. Their doxastic world will differ, too, and ultimately the identification is hindered. In this case, the narrative programs of the two subjects (their short- and long-term goals, and the means to achieve them) can differ. Furthermore, the performative level can be detached from the ethical level, in that the player will not be in a subjective position ('I shoot, I kill'), but rather on a third-person degree of responsibility ('My alter ego shoots, my alter ego kills').

A further element to take into account when discussing the distance between the user and the *simulacrum* is the dialectic between individuality and multiplicity. A peculiar aspect of computer games is that the alter ego need not be a singular character, but can be many entities at once, a group, a function in a system or even the whole system. This process of semiotic actorialisation can be distinguished into four categories:

1. Absent actorialisation. There is no specific actorialisation and no investment of new identity (an in *Tetris*, or pinball/card games). The interactive text does not create a separate fictional universe, but it is configured as a digital extension of the real world.
2. Individual actorialisation. An individual actor corresponds to the user. This is a one-to-one relation that facilitates the shifting of the user to the fictional identity into a possible universe whose values and ethics are to be known and confronted.
3. Multiple actorialisation. The user controls many different entities at once (i.e. war games) or switches from one entity to another (i.e. sports games). The fictional universe thus created can be explored from a number of perspectives and through the specific features of each actor. The absence of a one-to-one relation between user and *simulacrum*, however, prevents a deep transaction of pathemic and ethical relation between them. On the other hand, when the player manages a multiplicity of actors, his or her ethical domain shifts from an ethics of individual behaviour (actor-world) to the ethics of group behaviour (ethics among actors).
4. Super-individual actorialisation. The user's *simulacrum* is the whole fictional universe. The user is a self-adjusting system (i.e. *Sim City*) whose entities and characters (in *Sim City*, cars, citizens and so forth) are not controlled by the user, but are affected by the macro-level interventions of the user on the fictional world (e.g. to build a road, to raise a mountain). The user's alter ego is co-extensive to the fictional universe, and therefore the ethics of interaction are no longer defined by the relation between the user's alter ego and the world it inhabits, but rather by a self-contained, metalinguistic principle: the user is appointed with the role of establishing ethics in the fictional universe.

As I have tried to show, a thorough approach to ethics in interactive media should expand its discourse from the ethics of the game world to the semiotic construction of the *simulacrum*, the subject of ethics. The semiotic *débrayage* that transforms the user into a *simulacrum* is a highly complex phenomenon that employs a variety of semiotic devices (user competences, narratives, axiological systems, interface, point of view, singularity/multiplicity of the actor) to actualise this transition on the cognitive, pathemic and ethical level. In particular, the *distance* that separates user and *simulacrum* is a measure of the ethical responsibility of the former for her or his actions within the fictional world. To investigate the user-*simulacrum* distance means to venture into a field of study that requires a specific methodology and epistemology, a reworking of the theoretical apparatuses of semiotics, media studies, cyber-identity studies and narratology. The outcome of such an investigation could be a more profound understanding of what happens when a user enters a fictional world, takes on a new role, and plays, acts and performs within the domain of the ethics of interactive media.

11
'Moral Management': Dealing with Moral Concerns to Maintain Enjoyment of Violent Video Games

Christoph Klimmt, Hannah Schmid, Andreas Nosper,
Tilo Hartmann and Peter Vorderer

Introduction

Many of the most popular video games contain violence (for example, see Smith, Lachlan, and Tamborini, 2003). Players of violent games take the role of soldiers, policemen, secret agents or professional killers (*Hitman: Contracts*, 2004), which implies war action or 'small-scale violence', such as shootings or bombings. Rapid progress in computing technology has rendered the audiovisual appearance and the interactive quality of today's video-game violence extremely realistic, authentic and dynamic (Tamborini and Skalski, 2006). In fact, the simulation of violence in some video games has become so authentic that the military has discovered them as new opportunities for training and recruiting (*America's Army*, 2002; *Full Spectrum Warrior*, 2004).

The quantity and quality of video-game violence is being observed very attentively and critically by researchers, politicians, parents and public institutions because game violence has frequently been accused of facilitating the development of an aggressive personality. Research in psychology and communication has accumulated evidence for the existence of such an effect: massive use of violent video games causes an increase in aggressive thinking, hostile emotions and readiness to commit violent actions (Anderson, 2004; Sherry, 2001; Slater, 2003). Concern about the consequences of exposure to video-game violence is therefore well-justified.

However, little is known about the psychological mechanisms behind this effect, and also about why violent video games are extremely popular. A variety of theories have been advanced to explain the connection between game violence and aggression (Klimmt and Trepte, 2003; Weber, Ritterfeld and Kostygina, 2006). However, both the theories on media violence and the majority of the research conducted so far display at least two important shortcomings. First, they tend to ignore higher cognitive processes during the consumption of media violence. We do not know much about the way players think about the violence displayed on the screen or how they experience violent content (Potter and Tomasello, 2003). Second, most relevant research has neglected the fact that media violence is frequently embedded in entertainment contexts (Goldstein, 1998), for which violent video games are a prototypical example (see Jansz, 2005). If people use media violence to achieve pleasurable experiences, this motivation should be considered when theories on the effects of media violence are formulated, primarily for two reasons. One is the downward spiral problem (Slater, 2003; Slater, Henry, Swaim, and Anderson, 2003; Slater, Henry, Swaim and Cardador, 2004). If people find media violence enjoyable, they will consume it repeatedly, which increases the probability of effect intensification. The other reason is that video-game enjoyment is thought of as arising from highly complex cognitive processing (Klimmt, 2003). New and very specific psychological mechanisms that underlie game violence effects may be identified if the complexity of the enjoyment of violent media content is considered systematically.

The following theory thus relates to both the pleasurable qualities of video-game violence and its potential relevance for the impact of game violence. It addresses one important puzzle related to the construal of video games' appeal, *moral reasoning during game-play*. Violent video games offer players the opportunity to perform actions which contradict common moral standards, such as harming or killing virtual characters (Smith et al., 2003). In real life, intentions to commit violent actions are most frequently not executed, primarily because moral reasoning inhibits the realization of wishes to harm others or damage objects (Bandura, 2002). If violent actions are executed, however, moral ruminations often cause remorse as well as feelings of guilt and sadness (Düwell, Hübenthal and Werner, 2002).

This important role of moral reasoning for action regulation should also apply to players of violent video games. The execution of violence in games directly relates to moral reasoning, which is a threat to enjoyment – feelings of guilt and remorse would certainly undermine pleasurable experiences, such as pride or suspense (Klimmt, 2003). It is therefore interesting to analyze how players of video-game violence deal with moral concerns that are expected to arise from their activity. We introduce the concept of 'moral management' as a theoretical approach to this question. It applies Bandura's (2002) theory on moral disengagement in real-life aggression to video-game players in order to develop an explanation for the enjoyment of violent video games in spite of moral concerns. Moreover, the concept allows the researcher to identify a specific and complex psychological mechanism of game violence effects, which relates to the acquisition and rehearsal of strategies to cope with moral concerns when violence promises to be of instrumental utility.

Enjoyment of video-game violence

The literature on the reasons why people find the use of media violence, especially violent video games, entertaining is surprisingly sparse (Sparks and Sparks, 2000; Klimmt and Trepte, 2003). Based on findings from audience research that portray young males as the main user group of violent video games (for example, Roberts, Foehr and Rideout, 2005), Jansz (2005) argues that violent games are of developmental value for young male adolescents because they provide a testbed for the construction of their gender identity (see also Kirsh, 2003). The importance of violence for the definition and elaboration of a male gender identity is therefore considered as an important reason why game violence is appealing (at least for adolescent males).

Kuhrcke, Klimmt and Vorderer (2005) have proposed additional explanations as to why playing violent video games can be enjoyable. One important factor is the aesthetic pleasure of destruction. Referring to assumptions of Allan and Greenberger's (1978) theory of vandalism, they argue that there can be some aesthetic fascination in processes of destruction. For instance, observing a collapsing building or a man being shot in slow motion can evoke some aesthetic pleasure, which would contribute to the appreciation of an entertainment medium that displays such destruction. In addition, Kuhrcke, Klimmt and Vorderer (2005) have proposed that violent game elements may foster mechanisms of enjoyment that apply to all video games. Klimmt (2003) has argued that successful resolution of tasks and challenges during video-game play evokes pleasurable increases of self-esteem because players perceive themselves as competent and being in control. Such self-enhancement through game-play may be especially effective if the game actions involve violence (Kuhrcke, Klimmt and Vorderer 2005). Armed conflict creates high-stakes situations of competition ('dead or alive' confrontations) which should boost self-esteem especially strongly if a situation is mastered. Similarly, the possibility of committing violent acts (such as killing characters or destroying cities) is a clear and pure manifestation of control and power, which also evokes high self-esteem and a positive self-perception in players (Hartmann, 2003; Klimmt, 2003).

In summary, the available literature base suggests that game violence is enjoyable for several reasons. Among the most important factors is a game's capability of allowing the construction of a male gender identity and of creating circumstances under which situational self-enhancement based on success and power functions is effective. Consequently, a positive, proud and (super-)male self-image is proposed to originate from the entertaining use of violent video games.

Moral concerns: a challenge to game enjoyment

When the first game of the tactical combat simulation series *Hitman* was published, it stimulated a lively discussion in some game magazines. In contrast to most commercial games known until then, *Hitman* put players in the role of a professional killer whose tasks included murder of innocent (or at least, defenceless) individuals. Therefore, some journalists found the game morally inappropriate. The most recent sequel of *Hitman* attempts to avoid such moral troubles and identifies only 'bad guys' such as Mafia leaders as targets for the Hitman. This anecdote is interesting for the investigation of game enjoyment, because it suggests that the pleasures of using violent video games can be reduced or even diminished when moral justification for players' violent actions is absent.

Among the dimensions of enjoyment of violent media use (see above), moral concerns would probably not affect the pleasurable experiences related to the construction of a male gender identity (Jansz, 2005; Kirsh, 2003). Non-justified violence may be agreeable if a hyper-masculine gender ideal is pursued, which includes the perception of 'violence as manly' (Mosher and Sirkin, 1984). Hyper-male ideology, such as socio-evolutionary views ('only the strong survive'), may render morally inappropriate violence tolerable or even desirable. If the acquisition of a super-male gender role is the main driver of video-game enjoyment, moral concerns should therefore not reduce or eliminate the fun of playing.

The other proposed routes to enjoyment in violent video games are more susceptible to negative influence by moral concern, however. Aesthetic pleasures of destruction will occur only if the situation allows players to focus on the fascinating images and sounds of explosions, bullets, fire and so on. Moral ruminations, for instance, about innocent people caught in a collapsing building, would turn the observation of the destruction into a mixed-feelings situation (fun plus remorse) that would be much less enjoyable than a process of destruction that is free from such ruminations. Similarly, the self-enhancement derived from success and perceived dominance would lose at least some of its pleasurable quality if success was achieved by killing innocent people, for example. In summary, moral concerns represent a severe threat to the effectiveness of several hypothesized mechanisms of enjoyment in violent video games.

The enjoyment of playing violent video games is therefore considered to be in permanent danger of being reduced or eliminated by moral concerns related to players' violent actions. The danger is permanent, because moral concerns are not only related to extreme cases such as the '*Hitman* anecdote' reported earlier. In general, any act of violence that is executed, or has been executed, typically breeds moral conflict (Bandura, 2002). In real life, justified violence frequently leaves agents in moral rumination just as unjustified violence does (for example, police officers who have killed a perpetrator in self-defence; see Williams, 1999). Theoretically, each violent act, each killed opponent, each destroyed building, car or other object should evoke moral conflict in players of violent video games. Because violent video games feature extreme violence against people and objects (Smith et al., 2003), sources of moral conflict are ubiquitous and potentially very powerful.

The persistent threat that moral conflict imposes on the enjoyment of violent video games requires players to respond if they want to maintain their pleasure (see Zillmann's [1988] mood-management theory). Players need to find an effective way to counteract moral conflicts. Moral Management Theory argues that players perceive this desire to counteract moral conflict, that players can rely on a variety of cognitive operations to fight moral conflict successfully, and that most violent video games provide ample support to apply these strategies.

'Moral Management': disengaging moral concerns while playing violent video games

Moral Disengagement Theory stipulates a set of cognitive operations from a real-life context that players of violent video games can employ to resolve the problem of moral concern related to their violent actions in the game. These operations have mostly been introduced by Bandura's (2002) theory of moral disengagement, which addresses processes of moral disinhibition in perpetrators of real-life violence:

- Morally condemnable actions are legitimated by invoking a higher social norm whose accomplishment justifies violence (*moral justification*), for instance, the maintenance of security or the defence of democracy.

* *Euphemistic labelling* of the consequences of transgressive behaviour downplays the perceived effect of refutable actions (for example, by using the word 'neutralizing' instead of 'killing') and makes the reason for moral concern (from the agent's perspective) disappear.
* *Advantageous comparison* means that one's own behaviour is justified by comparing it with more condemnable actions of others.
* In the case of *displacement or diffusion of responsibility* the individual responsibility for violence is transferred to others (such as 'commanders'). Violent behaviour is interpreted as a necessary consequence of orders 'from above' or as a result of abstract social processes (for example, 'the system') that seemingly do not allow for blaming the individual for the violence committed.
* *Disregard or distortion of consequences* refers to the downplaying of consequences of violence to disengage internal moral standards (similar to euphemistic labelling; see above).
* In the case of *dehumanization*, targets of violent actions are declared to lack human dignity and/or ascribe bestial qualities ('they are like animals'). This way, dehumanization weakens self-restraints against cruel conduct and prevents serious moral concerns.
* *Attribution of blame* justifies transgressive behaviour by thinking that the target of violent action deserves nothing else but violence.

Moral Management Theory argues that the situation of playing violent video games provides the opportunity to apply one or more of Bandura's moral disengagement strategies and also enables an additional route to cope with moral conflict. We differentiate two basic strategies of overriding moral concern. One strategy refers to emphasizing the virtuality and as-if quality of the gaming environment. The other builds on characteristics of the game world and utilizes cognitive operations proposed by Bandura.

Anecdotal reports of player responses to the question of how they deal with moral conflict when they are using violent video games reveal that experienced players tend to highlight the differences between real-life aggression and game violence. In violent video games, no real people are harmed and no real objects are damaged. Blood, fire and debris are only pixelwork, and all experience is simulated and virtual (Klimmt, Nosper, Schmid, Hartmann and Vorderer, 2006). The non-reality status of video-game situations is used to explain why moral concern is not 'necessary', applicable or rational in respect to violent game actions. This strategy to avoid moral conflict makes sense only in playful contexts and thus is a specific operation available only to game players. Moral Management Theory therefore proposes the activation of knowledge about the difference between game violence and real-life violence as one general strategy to suppress moral conflict during gameplay. However, the observations of video-game players who are feeling Presence (Tamborini and Skalski, 2006; Schneider, Lang, Shin, and Bradley, 2004) suggest that this strategy can probably not resolve the problem of moral conflict in all conceivable gaming situations, because players are presumably not aware of the game-reality difference when they are highly involved or absorbed by the game (Klimmt, 2003).

Therefore, Moral Management Theory argues that players can also deal with moral ruminations without focusing on the game-reality difference. Players can also rely on (adapted) moral disengagement strategies known from real life (Bandura, 2002) to override moral concern and maintain enjoyment. For instance, most violent games construct the player role as a brave hero who attempts to protect the weak (moral justification) on orders from kings or other mighty institutions (diffusion of responsibility), or who is fighting against ugly monsters (dehumanization) that are forcing the player to defend him-/herself so that they only deserve to be killed (attribution of blame). Alternative combinations of moral disengagement strategies apply to other violent video games.

Very different patterns of implementation can be imagined for these strategies. For instance, players may use the narrative framework outlined in the introduction episode of a new violent game to derive moral justification for all their violent actions that will follow. This way, they would 'switch off moral concern' in advance, that is, before the first violent act is committed. An alternative way to implement moral disengagement strategies during game-play would be to activate specific combinations of strategies for each case of violent activity. In role-playing games such as *World of Warcraft* (2004), conflict is not the only mode of social interaction. It is therefore reasonable to treat each character that appears on the screen individually rather than setting off any moral reasoning once and for all. Each conflict that arises in role-playing games may be paired with a specific combination of moral disengagement strategies. With increasing experience, categories of characters may be associated with defined moral disengagement operations (for example, 'monster type A deserves killing' = attribution of blame; 'monster type B has been condemned by the king' = diffusion of responsibility), which would result in very quick suppressions of moral conflict that do not consume much cognitive effort.

In sum, Moral Management Theory argues that (one or several) moral disengagement strategies outlined by Bandura (2002) are adopted by players of violent video games. This is the second general strategy of moral management that does not rely on game-reality differences, but even operates when a game is seen as 'real' (see the next section for details). As a conclusion for this section, the key proposition of Moral Management Theory is that players respond to the threat to the entertainment experience caused by possible moral concern by either increasing the salience of game-reality differences or by applying one or more (modified, specified) cognitive operations that have been outlined in Bandura's Moral Disengagement Theory.

The differentiation between the actualization of game-reality differences (that is, the activation of a distanced and critical mode of experience) and the application of moral management strategies are not considered as mutually exclusive. It is more plausible to assume that players will – in certain situations – combine both routes towards the override of moral conflict and will in other situations rely on only one of these routes. Because the game experience is highly dynamic (for example, suspense, success and death/failure can occur within a few seconds), strong temporal variations in the salience of game-reality differences and the perceived viability of specific moral disengagement strategies are also expected.

'Moral cues' in violent video games

Most violent video games frame the events that occur during their use in specific ways in order to enable and support players to cope with moral concern. Because the experience of Presence (non-mediation) is important for game enjoyment (Tamborini and Skalski, 2006) violent games do not emphasize game-reality differences (as one possible route to moral management). This way, the game would risk sacrificing some of its entertainment value. Rather, violent video games provide 'moral cues' to stimulate or enhance moral management strategies that function under the condition of Presence in the game world. For virtually any of Bandura's moral disengagement strategies, specific cues can be built into a game world that guide players' information processing. Consequently, players are more likely to discover and execute suitable and effective moral management strategies because they need only to respond to helpful game elements:

- As mentioned earlier, most violent games lay out a narrative framework that provides *moral justification* for violent action, most typically good-versus-evil stories that assist players to cope with ruminations about the sense and appropriateness of violence ('defend freedom', 'protect the weak').

- As the majority of violent games utilize military contexts and scenarios (Kuhrcke, Klimmt and Vorderer, 2005), typical strategies that commanders apply to stimulate moral disengagement in their troops, such as *euphemistic labelling* of violent action ('eliminate the enemy'), can also be found frequently.
- The worlds of most violent video games are full of creatures that perform extremely crude violence. They serve several functions in respect to morality, such as 'attribution of blame' (see below) and the creation of a projection screen for *advantageous comparisons* of violence committed by players. Compared to these creatures' brutal behaviour, players' violent actions are framed as 'moderate' and 'not exaggerated'.
- Violent games that feature military, police or intelligence contexts frequently provide narrative cues for the implementation of the *diffusion of responsibility* strategy, as they introduce higher authorities (such as government officials) that demand the execution of violence and/or free players from the perceived responsibility for violent incidents.
- Cues that refer to the *disregard of consequences* occur in virtually any violent video game at the narrative level (for example no widows or orphans appear after an opponent is killed).
- *Dehumanization* of victims is also facilitated by most violent video games. For instance, some games (*Resident Evil*, 1996) introduce monsters and other fantasy creatures that differ in some respect from humans. More implicit dehumanization cues build on outsider group perceptions and stereotypes (for instance, by designing opponent characters to meet the 'Arab terrorist' cliché).
- Support for the *attribution of blame* strategy of moral management is also typical for many violent video games, most often through creating situations of self-defence. Game characters that attack the player (or her/his) character obviously deserve violent treatment, as do victims who 'deserve' punishment, as they are themselves perpetrators of unjustified violence.

In sum, most video games that allow or even demand violent action come with a large number of cues that help players to select and execute moral management strategies successfully. By providing cues for moral management, games pave the way for enjoying violent action without guilt and remorse. The combination of possibilities to commit violence and disinhibition of moral conflict results in optimal entertainment experiences for those players who are able and willing to adopt the perspectives on morality that the game suggests.

Instrumental coping with moral concern as an effect of game violence

So far, we have outlined a theory of enjoyment of violent video games that is enabled by cognitive prevention or override of moral conflict. Moral Management Theory is therefore a process theory of the consumption of violent video games and provides an explanation for the formation of entertainment experiences during interactive media use. The theory also holds implications for the conceptualization of the effects of violent video games on aggressive thinking, emotions and behaviours. Research findings on the impact of violent games converge toward considerable effect sizes (Anderson, 2004; Sherry, 2001). However, the psychological mechanisms behind these effects are still in need of theoretical explication (Klimmt and Trepte, 2003). Moral Management Theory may help to discover one of these mechanisms.

The theory argues that cognitive operations to cope with moral conflict occur frequently and in highly diverse manifestations in players of violent video games. Moral management is expected to be an integral component of players' processing of game events. We therefore propose that the use of violent video games functions as a cognitive training ground for moral disengagement strategies that can also influence behaviour outside of gaming situations. Playing violent video games enables, first, the acquisition of different strategies (that is, the internalization of strategies in the behavioural repertoire), second, the rehearsal of quick and effective selection of possible strategies (that is, a 'suitable' moral disengagement strategy comes to individual's mind more quickly and more easily), and third, the effective handling of strategies in the execution stage (for instance, learning to ignore situational cues that could indicate the inappropriateness of a given strategy). These 'skills' would also be instrumental for moral disengagement in real life. Such transfers of moral management skills from game-play to real life could operate, for instance, via social-cognitive processes (Bandura, 2001) that integrate moral management operations into a player's general behavioural inventory and/or affect morally relevant attitudes (for example, Eyal et al., 2006).

A second theoretical account that can explain how the acquisition and rehearsal of moral management can transfer to real life is priming (Anderson and Dill, 2000). Through frequent activation and actualization during video-game play, cognitions related to resolution of moral concerns (including predefined strategies of moral management) become necessarily more salient in players' minds.

Consequently, these cognitions are easily accessible ('top of mind') for players in situations of real conflict. Heavy users of violent video games should thus display a higher cognitive 'preparedness' for overriding moral concern and thus for executing violent behaviours in real life. This contention does not imply, of course, that effects of moral management during video-game play are proposed as the new key explanation for youth violence. Rather, the concept should be regarded as an attempt to identify one complex and specific mechanism that links the consumption of media violence to aggressive cognitions, emotions and behaviour and that functions within the orchestra of other factors (most of which are not related to media consumption at all). Its focus on the enjoyment of violent video games marks a conceptual innovation to both the questions of the motivational attractiveness of violent video-games and the mechanisms underlying the association between violent video-game play and aggression.

12
Beyond Good and Evil: The Inhuman Ethics of *Redemption* and *Bloodlines*

Will Slocombe

'By becoming a monster, one learns what it is to be human.'

(*Vampire – The Masquerade* 1)

Studies of the ideological frameworks and ethical systems of computer games take many different forms, from analyses of how games promote certain types of emotional response to what kind of ideological factors realise themselves in the game world, but all are predicated on the fact that there is always an ethical system implicit to a game's architecture. Friedman (1999), for instance, focuses on the specific manifestations of ideology in the games he studies (such as pollution in *Civilization II*, 1996). Other theorists, such as Sicart (2005), take a broader view, studying the ways in which ideological and ethical beliefs are coded into games. This essay takes a slightly different approach to the study of ethics in computer games because its subjects, *Vampire – The Masquerade: Redemption* (2000) and its sequel *Vampire – The Masquerade: Bloodlines* (2004), are self-evidently works of fantasy. Each might easily be dismissed as yet another game that has no bearing on real life, although this misses the fact that there is still an ethical system embedded within both, an 'inhuman' ethic that is 'beyond good and evil'.

Redemption begins in medieval Prague and the avatar, Christof, is a Crusader. Tortured by his own spiritual doubts and his love of a nun, Anezka, Christof is turned into a vampire. While he continues to fight vampires, he is now part of that society and so seeks to kill as many 'evil' vampires as he can find at the behest of 'good' vampires. The arc takes the player through to New York in 1999 and the awakening of an ancient vampire, Vukodlak. There are no choices during character creation, but there is room for character development in what powers the avatar can learn as the game progresses. In comparison, *Bloodlines* is a less narrative-driven game and, at least initially, more open. Where *Redemption* allows the player to begin playing as a mortal, *Bloodlines* begins with a cutscene in which the avatar is 'embraced' (that is, turned into a vampire). Players select which clan (a particular bloodline of vampires) they join and their initial statistics. Located in contemporary Los Angeles, from this point on the player is immersed in the politics of vampire society in a more explicit mission-based structure. Although there is still an overarching narrative, the quest to discover whether a sarcophagus holds a vampire of unimaginable power, the game is episodic, and more of a first-person shooter than a role-playing game with game strategy overshadowing character development.

The inhuman ethic behind the games is evident here – *Redemption* and *Bloodlines* have players assuming the role of vampires. Although all role-playing games encourage the player to take on a 'role', these games use the mechanics of an earlier game that explored human morality, White Wolf's Storytelling system. In comparison to the *Baldur's Gate*, *Neverwinter Nights* and *Icewind Dale* games, which are derived from *Dungeons & Dragons* and its quest-based, combat-heavy system, the Storytelling system focuses on the morality of player choices. *Vampire – The Masquerade* (*VTM*), the game upon which *Redemption* and *Bloodlines* is based, proposed that 'By becoming a monster, one learns what it is to be human' and thus from the outset declared its emphasis on morality.[1] *Redemption* and *Bloodlines* are therefore predicated upon a pre-existing ethical system and the problem with conversion to a digital environment is that much of this emphasis on morality is lost.

Storytelling the 'World of Darkness'

In *VTM*, players create vampire characters and act out parts in stories that dwell on the darker side of human nature; the Storytelling system makes players play the monster. These are not, however, 'evil' games: 'Though our purpose is not to offend, our use of vampire as a metaphor and as a channel for storytelling may be misconstrued. To be clear, vampires are not real. The extent to which they may be said to exist is revealed only in what they can teach us of the human condition and of the fragility and splendor which we call life' (*VTM* 2). The purpose of playing a vampire is to explore what it means to be human, to step outside socially circumscribed moral parameters and explore the nature of good and evil. The setting for this game is the 'World of Darkness', a much bleaker version of contemporary society in which monsters exist, but hide their existence from humanity. Furthermore, players do not play *VTM* to win, but to explore this world and their place in it.

Thus, *VTM* is concerned with social interaction and morality rather than a straightforward path to victory. The game arbitrates morality by means of a sliding scale of humanity, from the most saintly acts (rated at ten) to the most depraved (rated at one); the human average is a humanity score of seven. As players progress through the story, their avatars gain and lose humanity by way of a 'Hierarchy of Sins' (*VTM* 200–1). A 'sin' in this context is the most immoral action a character of a given level of humanity will perform. For example, characters with a humanity of six can commit theft without it affecting them (theft is only a sin for those with a humanity of seven or higher) but they cannot commit wanton destruction (the sin of a humanity score of five) without being forced to check if they have lost some of their humanity. This check forces players to consider their actions carefully, because the aim of the game is to hold onto what little humanity they have in the face of their lust for blood. When the avatar's humanity drops too low, the player is removed from the game: her/his character is little more than a blood-hungry beast. This approach to ethics is not culturally relativistic, due to the Christian morality implicit to the *VTM* mythology: all vampires are descended from Cain, the brother of Abel and the first murderer.[2] Called Caine in *VTM*, he is the progenitor of vampires, cursed to wander eternally. Taking as its basis the Ten Commandments, the 'Hierarchy of Sins' thereby embeds Christian ideology within *VTM*. However, this system translates badly into a digital environment and both *Redemption* and *Bloodlines* demonstrate the extent to which the perceived needs of gamers have affected what was intended as an exploration of human morality.

In *Redemption*, for instance, killing a human results in a loss of five points of humanity (it measures humanity as a percentage). Rather than the 'Hierarchy of Sins' of *VTM* (which actually *allows* a character to kill if s/he is of sufficiently low humanity), *Redemption* referees ethical actions simplistically and, arguably, leniently. There are also occasions when dialogue choices affect the avatar's humanity score. These are in some ways better indicators of a player's ethical position, although such options are rare and have relatively little impact on game-play past the fact that, as with *VTM*, *Redemption* ends when the avatar reaches a humanity score of zero. One particular ethical dilemma is the killing of Luther Black, a vampire who wishes to suffer Final Death:

> Luther: If thou would redeem, redeem me.
> Christof: I shall not kill thee.
> Luther: Please! I beg of thee! I cannot enter Heaven if I turn my hand against myself. But thou … thou has it in thy power to deliver. 'Twould be an act of holiness.

Christof: If I take up thy burden, thou wouldst remain blameless and thy sin would be mine alone to bear. Thy elevation would be my damnation!
Luther: Nay.
Christof: Aye, 'tis so.
Luther: Aye, 'tis so. Wilt thou do this for me all the same? It is more than I may ask of thee, yet I ask it still. Wilt thou take my sins upon thee?

If the player decides to help Black die, the avatar takes on Black's sins. Those players who have explored this option note that this means they can only get the worst ending of the game, the one that relies on the least humanity. While killing an evil vampire might be deemed good, the morality revealed here is that individuals must look to preserving the state of their own souls.

Bloodlines also uses the humanity system, this time scored zero to ten, albeit with some differences. There is still a fixed penalty for killing humans (one point, with a lowest score of three) and dialogue options still affect it. The major difference, however, is that humanity changes from an objective measure of the avatar's humanity to a relativistic 'being the person you once were' (*Bloodlines* manual, p. 20). It is also vital to observe that the game-play effects of humanity are also lessened: in *Bloodlines* it is impossible to reach a humanity of zero and it influences game-play only inasmuch as it makes the avatar more likely to 'frenzy' (the player loses control of the character). *Bloodlines* also introduces another concept, the 'Masquerade violation'. 'The Masquerade' is the vampires' method of hiding themselves from humanity, their social compact to ensure that vampires remain hidden from a human society that is capable of eradicating them.[3] This is dramatised within the game through three kinds of environment: combat zones, Masquerade zones and Elysium zones. Elysium zones do not allow avatars to attack, drink blood or use special powers, whereas combat zones allow all three.[4] Masquerade zones are the middle ground between these, and most simulated public areas in the game are configured in this way. Although Masquerade zones allow combat and the use of disciplines, there are consequences if the avatar is seen using particular powers or attacking someone. Attacking or feeding provokes a hostile response from guards or police, like *Redemption*, but a unique feature of *Bloodlines* is that using disciplines in such game areas invokes a 'Masquerade violation'. If the avatar is witnessed using special powers five times, s/he is killed by the other vampires. There are in-game opportunities to gain 'redemptions', but most players will be careful where their avatars use their special powers.

The blood that moves the body: ethics and interactivity

The inclusion of 'humanity' in both games, and 'Masquerade violations' in *Bloodlines*, means that a certain level of ethical behaviour is determined by how the player interacts with the game environment. While these are obviously numerical indicators that affect game-play, they are only a crude measure of the ethical path that the player has taken through the game. How the games score these 'ethical' decisions is imbalanced, for they neglect a wider context to avatar actions as often as they ensure that some player decisions, which are themselves ethical, are not placed within the context of humanity.

As Reynolds (2002) notes, however, there are alternative methods of reading games. For example, applying normative ethics (what is deemed 'normal' within a given community) allows us to see that 'Masquerade violations' are in fact the normative ethics of vampiric society. Thus, *Bloodlines* actually dramatises the conflict between a humane 'humanity' and an inhuman(e) 'Masquerade'. This is seen clearly in certain quest resolutions in *Bloodlines* where players must choose between losing humanity or gaining a 'Masquerade violation': if you want to be a 'good' vampire (in normative terms), you must be a 'bad' human. In the 'Carnival of Death' quest, for example, players try to deduce who committed a number of grisly murders. When the perpetrator is caught, the player has two choices, either to persuade the killer not to kill again (he was killing 'evil' humans) or to kill him. The first choice gains the avatar a point of humanity – presumably because of empathy – but also a 'Masquerade violation'. The second solution, the killing of a 'justified' killer, offers a 'Masquerade redemption' instead. 'Attention Whore' works similarly, as a vampire groupie must be stopped from breaking the Masquerade. One solution is to persuade the groupie that her vampire idol has skipped town – she then leaves Los Angeles (which shifts the problem rather than solves it), at which point the avatar gains a 'Masquerade redemption'. The second solution involves killing her. The player may decide either to kill the groupie or to send her to a vampire that eats human flesh; both involve a loss of humanity and a 'Masquerade redemption'. This choice also appears earlier in the game, when a reporter uncovers too much information about vampires. Here, sending the individual to the flesh-eating vampire rewards the avatar with a 'Masquerade redemption' but no loss of humanity. This is evidently disturbing, for according to the game mechanics it is sometimes justifiable to intentionally send someone to their deaths. While there are often ways around these quests (an example of gaming logic over narrative consistency), the solutions elide very real moral dilemmas. Although it pretends to dramatise ethical choices, *Bloodlines* avoids the problematic morality of the 'World of Darkness'.

This tendency to avoid ethical dilemmas in favour of simplistic numerical reductions is what makes *Bloodlines* such a deeply unethical game. You may kill those responsible for snuff movies and avenge injustices, but often the solution is still to kill. Moreover, one of the staples of vampiric existence – the potency of blood – is dealt with in a very off-hand manner. To explain: vampire blood is addictive and, when fed to mortals, makes them into ghouls – humans who can live forever providing they regularly feed on vampire blood. The groupie mentioned above is one such ghoul in the game, and thus the player's potential decision to kill her solves a problem created by another vampire in the first place (it is not the groupie's fault she desires the vampire, after all). A second example is the ghoul that the avatar can create. In the first section of the game, the avatar sees a dying woman in a clinic. The player can choose to give her some blood (which will heal her) and thus gain a point of humanity. This act makes the woman into a ghoul, bound into servitude, although the avatar can gain a point of humanity by rejecting her. If the player accepts her as a ghoul (with no humanity loss!), as the game progresses she morally deteriorates, seducing and kidnapping a victim for the avatar. There is no 'Masquerade violation' solution to this interaction: the player can only release the victim by threatening him with violence or accusing him of attempted rape, otherwise the only choices are to use a power that will lead to him getting himself killed ('You should go to the cops right now. Tell them "I've got a gun, pig-gies"') or to kill him. Should the player decide the latter, the ghoul helps dispose of the body, thus turning a college student into an accessory to murder. To make matters worse, such behaviour is rewarded by the game as the ghoul provides the avatar with the best armour. Although afterwards the ghoul is killed, most players will care only about the armour.

The game-play around the ghoul is also suspect and indicates a serious ethical problem in *Bloodlines*: its representation of women. Your ghoul can be assigned a choice of three outfits, all of which are presumably designed to be sexually provocative. This is illustrative of the game as a whole, with women being little more than sexual objects. For example, the avatar has the opportunity to sleep with different characters in the game. While these are not graphically depicted (noises only), there is a misogynistic bias here because it is perfectly possible to sleep with all of them if the character is female, but only with women if the character is male. All of the prostitutes and 'blood dolls' in the game are likewise female and can be seduced by characters of either gender, although *Bloodlines* cannot be described as enlightened towards homosexuality as it is rather a voyeuristic male fantasy. The metaphor of vampire as sexual predator here represents phallocentric sexuality, which explains the quest for four 'glamour' posters of significant female characters in the game and the existence of a cheat, 'money x', that affects the breast size of all female models. Even the character dolls reveal a certain perception of women. During character creation, the majority of female models are of the same type – thin, attractive and young (with the exception of the repulsive Nosferatu, but they are still thin and young) – whereas male models demonstrate much more individuality, especially the Malkavians (insane vampires).

'It is the blood of Caine which makes our fate': predestination and linearity

The endings of the games also reveal the extent to which game-play has affected any exploration of morality. In *Redemption*, for example, despite its three different endings, there is not one that could be called 'human', a fact equally true of the various endings of *Bloodlines*. This may be partly due to verisimilitude – after all, the avatars *are* vampires – but such endings nevertheless reduce the ability of the games to promote moral awareness.

The endings of *Redemption* are tied to the avatar's humanity at the close of the game. As has been previously mentioned, a low humanity is dangerous for players as a humanity of zero causes the game to end as the avatar succumbs to the Beast within. Nevertheless, it is possible to finish the game with a low humanity score. When players do this, the final fight with the vampire antagonist, Vukodlak, results in only one possible conclusion: the avatar feeds on Vukodlak's blood, taking his power. Here, the avatar no longer cares about mortals and seeks to rule the world, forsaking his allies and killing Anezka. The second ending is only possible on a relatively low level of humanity. In this ending, the player can submit to Vukodlak, who then has the avatar kill Anezka. It is the final ending that is most problematic from a moral perspective, however, as an avatar with their humanity relatively intact (60-plus) still behaves inhumanly. As with the previous endings, the player must win combat against Vukodlak, but in this ending, there is a further conflict. Vukodlak banishes the player to another area where they encounter Anezka's 'wall of memory', revealing her desire to see Christof again. After defeating Vukodlak again, there is a cutscene in which Christof 'saves' Anezka from death by 'Embracing' her (her existence was dependent on

Vukodlak's blood). In this ending, it is clearly not 'redemption' (as the title suggests) but instead the acceptance of damnation, as Anezka states: 'Damnation with you is as sweet as salvation'. Thus, the illicit love between nun and Crusader finally results in eternal damnation for both as they become vampires, albeit more 'humane' vampires than the antagonists throughout *Redemption*.

Bloodlines offers a variety of endings as well, only loosely connected to the game-play that precedes them (and, unlike *Redemption*, humanity has no bearing on the ending). In these endings, all of which revolve around the mysterious sarcophagus, the player may side with various factions and either 'live', get blown up, or be sunk to the bottom of the ocean. Nevertheless, all but one of these endings have the same generic cutscene, and it is this that demonstrates the inhuman ethic of *Bloodlines*. Smiling Jack, who aids the player throughout the game, talks to the taxi driver who drove the avatar to their final mission. They are watching the explosion of a skyscraper, caused by Smiling Jack when he substituted the body interred within the sarcophagus with explosives. This conversation contains one of the most significant lines in the entire game: 'It is the blood of Caine which makes our fate.' As mentioned earlier, in *VTM* Caine is the father of all vampires and so this statement is in many ways a truism; all vampires in the game are connected to Caine by their blood and, as they are vampires, Caine has evidently made their fate. However, the game folder contains a sound file of dialogue for the taxi driver, labelled 'Caine'. This brings the player *out* of the game into the linearity with which they have been involved. To explain this, 'Caine' is an allegory for the computer behind the game. It determines which of the possible endings the player chooses (most of which conclude in the same manner anyway) and throughout the game, when the avatar frenzies, the computer – 'the Beast' – takes control. This can of course be overstated, but just as the ethics of the avatar are determined by their vampiric nature, so too are the ethical actions of the player *in the game* determined by the way in which the game is set up (precisely Sicart's (2005) point about the ethical architecture of games).

Both *Bloodlines* and *Redemption* (but primarily *Bloodlines*) therefore come dangerously close to being inhuman and the danger of such games is that, as Nietzsche (1990) wrote, 'when you gaze long into an abyss the abyss also gazes into you' (p. 102). Looking at another ten commandments, written by the Computer Ethics Institute rather than Moses, the shortcomings of *Redemption* and *Bloodlines* become clear: '9. Thou shalt think about the social consequences of the program.' Even if we posit that the inhuman focus of both games is a partial attempt to deal with this issue, much of the implicit ethical programming is nevertheless disturbing. In *Bloodlines* especially, there is a clear lack of consideration of the social consequences of the game and, in both games, their inhuman ethics take us dangerously close to missing what it is to be human.[5] This statement is perhaps harsh, especially in relation to *Redemption*, because at least that game tries to follow the original idea of *VTM*, 'By becoming a monster, one learns what it is to be human'. Because *Redemption* includes a Prelude (a staple of *VTM* that deals with the character's life before the Embrace) and humanity determines the ending of the game, it is partly concerned with what it means to be human. We might not actually learn what it is to be human from *Redemption*, but at least we learn what it is to be an ethical monster. Santos and White's conclusion about survival horror games, 'the "evil" we encounter in these games represents the fragility and duality of our own psyches' (2005, p. 77), is equally applicable to games such as *Redemption*, for we can at least begin to appreciate the moral axis of our actions. This awareness is accomplished by *Redemption* primarily because of its explicitly linear structure and emphasis on humanity as an interactive tool. Unfortunately, *Bloodlines* is a world of ethical darkness where players cannot hope to escape the morass of evil, torture and murder. By removing the ethical framework from *VTM* to make *Bloodlines* more action-oriented, by making humanity a number to calculate frenzy rather than something that judges morality, and by implicitly promoting a socially unacceptable position throughout the game, *Bloodlines* does more harm than good even if we try to imagine that it has no social implications. The avatar may be beyond good and evil, but players are not: they are forced to behave inhumanly by the very nature of the game and this is no longer gazing into the abyss, but playing inside it (the only 'ethical' choice here is to not play the game). As the flesh-eating vampire in *Bloodlines* states, effectively summarising the game itself: 'I have no interest in morality. Only cause and effect'.[6]

While programs can only ever be 'cause and effect', the games market has developed sufficiently to allow games to be both enjoyable (else they are not games) and ethically stimulating (else they may be inhuman games). Games such as *Black & White* (2001) and *Fable* (2004) are still simplistic in moral terms but at least they are game-based explorations of ethics; unfortunately, neither *Redemption* nor *Bloodlines* lives up to *VTM*'s legacy.

Notes

1. As Halberstam notes in *Skin Shows* (2002, p. 37), however, the dichotomy between monster and human is far from straightforward: 'By demanding that

the monster round out our definitions of "human" (either by representing a polar opposite or by showing "real humanity") we also remake the monster as alien, as other, as difference. The monster, in fact, is where we come to know ourselves as never-human, as always between humanness and monstrosity'. That is, monstrosity is not diacritically opposed to humanity; rather, the monster's purpose is to reveal the extent to which humans are *already* partly monstrous. This is why this essay refers to 'inhuman ethics', for the ethics embedded within these games do not reveal facts about monstrosity, but about inhumanity.

2. Various versions of this link between Christian mythology and vampires exist. Alongside Lussier's rather prosaic *Dracula 2000* (2000), in which Dracula is Judas Iscariot, one might also think of Coppola's *Bram Stoker's Dracula* (1992) in which Dracula curses God after the death of his wife, or even Joss Whedon's *Angel*, about a vampiric 'messenger' with a soul.

3. This description of the Masquerade refers to the first edition of White Wolf's game, not the new edition covered by the *World of Darkness* rulebook, as the first edition is the basis of *Redemption* and *Bloodlines*.

4. Note that combat zones are one of the most ethically dubious areas of the game as they often allow avatars to kill innocent humans with impunity. For example, when the player first visits the internet café in Hollywood, it is a Masquerade zone. The second time, however, it has become a combat zone and so avatars can freely kill any humans that are still there. This is evidently a problem of coding ethics into 'environments', rather than in what might be termed 'entity triggers'. *Bloodlines* solves this problem by forsaking moral judgments on certain actions, evidently altering *VTM*'s very deliberate attempts to place players in morally ambiguous situations.

5. Of course, it is possible to argue that by showing players how *not* to act, the games promote humane actions. This seems improbable, however, precisely because of these games' reductively numerical morality; there is no complex moral code embedded within them. We might therefore argue that morally ambivalent or abhorrent actions are not the product of target dehumanization, but 'internalised inhumanity': avatars are not meant to be concerned about humanity and so, by extension, neither are their players.

6. Being cynical, it is worth noting another in-game quotation that sums up *Bloodlines*. Given the unpolished nature of the game, with numerous grammatical and spelling errors, and bugs, perhaps what the fortune-teller in Santa Monica tells the avatar is of more than humorous import: 'Whether or not you win the game matters not. It's if you bought it.'

Part V
Politics and Ideology

13
Preconscious Apocalypse: The Failure of Capitalism in Computer Games

Sven O. Cavalcanti

All forms of games have, unless played by the ruling class, always been a noncommittal modification of reality with the sole purpose of providing pleasure. In games, triumph is possible even for those whose actual life is no bed of roses. An old German saying is 'Aus Spiel wird Ernst', meaning that in the serious player's mind the boundaries between playing and real life are blurred. Gaming has never been on par with reality but has, at the same time, never been so far from reality that a truly free life could have been derived from game-playing: every game has rules which are deduced from economic, cultural and historic moments. In Germany, Ludo is called Mensch-Ärgere-Dich-Nicht ('don't get angry, man'), where the title already announces what the simple life is all about. Monopoly allows the player to climb to the position of a hotel magnate. And in *Risk* you may conquer the world.

Nevertheless, using the example of games, a relationship between the individual and social reality can be deciphered. Gaming does, however, also imply a flight from reality into a dream world of supposed equal chances, equal entry into a world of daydreams.

Most computer games are set in a world that has to be understood as an extension of present societies. Many of them visualize limbos, failed societies in which the player is provided with the role of an omnipotent saviour. The ability to engage in a game's lost worlds represents the abortive psychic dispositions of capitalism. By identifying with an avatar, the subject is a consumer of her own failed future, her own nightmarish life. The alleged fight for a better world that appears at the game's end and at the same time serves as a layer of legitimation for the game's slaughter worlds does, however, originate in the player's adjusted, one-dimensional consciousness. The fear of being smothered in existing societies promotes forces the individual wants in order to cope with her own life. At the same time, the failure of contemporary major topics is accepted and anticipated.

Thus, computer games are nothing but a subtle, sensual variation on completely ordinary games. Still, they are equally able to draw on the stylistic devices of film and transform the player into an interactive actress who accepts her role. This article will focus on those aspects of computer games that anticipate the failure of individual and collective designs in virtual game worlds – those computer games in which the world seems to be apocalyptic, an inverse foil of modern societies. The emphasis here will be on the gloomy realms of utopia that spread their obscure atmosphere in the game.

In order to make these counter-utopias playable, they need a preconscious moment in the player's mind: the unconsciously anticipated possibility of the failure of capitalism. Consequently, these sepulchral computer games appear to be unintentionally enlightening. Like no other cultural domain, computer games insist on the world being a hostile place that is by no means a favourable location to live in – nowhere else is the future definitely labelled as abortive. The subject's isolation in modern societies establishes her relationship to the limbos of failed computer game worlds; collectively and internationally, gamers play the same nightmare. In computer games, exit is a theological motif: deliverance from the evil on earth, from the evil in the game. Being a hero in computer games as well as in movies provides the very omnipotence that the drab monotony of everyday life lacks. In other words, the computer game is about a lost world, not about an averted world. It displays nightmare fantasies that already exist.

Economic catastrophe – the reign of rackets and gangs, social misery

A basic element of contemporary first-person shooters is the construction of a failed political world out of which modern capitalism's regalia can still be read. The world is ruled by rackets, with the population simultaneously impoverished. The gloomy interworlds of *Deus Ex* (2000) tell a tale about a cyborg's morality, and *deus ex machina*, the god in the machine. The battle between Illuminati and the *WTO* stands as a symbol for the battle between sheer market principles and religious delusion. Admittedly, *Deus Ex* actually includes critical elements since the player is allowed to decide between a dictatorially adjusted market and a religious delusion, and in the end makes clear that neither of these positions leads to a better society. None of them delivers salvation – only the sequel to 'the same old story' – and just another domination of another faction.

In addition, the existence of slums, ghettos and poverty belongs to the repertoire of game worlds that deal with economic catastrophe, such as *Max Payne* (2001) or the *Deus Ex* series that presents a descent into poverty. The avatar has to go to gloomy cesspools in order to ascend to heroism in the game's cathartic end. Social explosives meet their counterpart in rocket launchers by means of which evil dictators or racketeers will be blasted. The Marxian alienation principle also finds its expression in the *Lumpenproletariat*[1] in the form of rival gang fights for territory and profit. The *GTA* series does, on the one hand, glorify the application of violence (like almost any computer game), but on the other hand exaggerates and caricatures the depiction of those cities where the avatar engages in its bloody bleakness. The actual development of a ghetto experiences its disassociation in *totalisation*.[2] The police are reduced to corrupt animals for slaughter and do not lay a moral claim on the enforcement of civil or human rights. This vision of a ghetto city with no rules is a sardonic criticism aimed at capitalism. The welfare state principles have been scrapped – instead, psychically deformed dropouts battle for existence. The bizarre forms of reality that *GTA San Andreas* (2004) has adopted can be imagined as a flooded *San Andreas* – how far would this imaginative vision be from the pictures of New Orleans in the aftermath of hurricane Katrina?

Economic catastrophe – the reign of dictatorship

Another scenario consists of the formation of political dictatorships after economic catastrophes. An avatar in *Half-Life 2* (2004) surveys a depressing world where a fascist military dictatorship oppresses the population and the paramilitary police execute their tyranny. The shadow realm's luridness in *Half-Life 2* is perfected by its brilliant narrative style. The game's main story spares briefings and previous histories; the player is immediately confronted with a reality of oppression and arbitrariness in which she almost instantaneously finds her way and intuitively joins the resistance movement. At the same time, *Half-Life 2* deals with a 'placeless' surveillance state. The architecture is plain and may find its equivalent in every halfway modern city – police uniforms match contemporary operation units' uniforms. It hardly matters where on this planet *Half-Life 2* is set. The evocation of a depressing reality is possible because almost every country has experienced excessive task force missions.

Additionally, anti-Semitic stereotypes in computer games have increased. In *Under Siege* (2005), Syrian programmers coded a scenario which not only questions Israel's right to exist but in which the (virtual) killing of Jews is the basis of the legitimacy layer of violence. This signifies a new quality of computer game politics. Here, more than just the battle against occupying forces is being evoked. Games such as *Special Force* (2003) can be described as a hidden appeal to destroy Israel itself; the anti-Semitic resentment evolves into heroism. Without recourse to economic catastrophe and the failure of capitalism, however, it cannot be understood why people are willing to be degraded to goods or to kill pixel bodies. Computer games are an image of the era and social order they come from – the social deadness of late capitalism not only shows up in reality but also in the cultural assets it produces.

Economic catastrophe – the reign of war

Corresponding to the failure of the contemporary economy are those games where nation states irreconcilably oppose each other. World war scenarios, particularly of the First or Second World War, fall into this category – games in which war is absolute and the player shares the soldier's virtual fate. In *Operation Flashpoint* (2001), the staging as an anti-war game fails, but the player loses her understanding about the reasons for war. In *Codename: Panzers* (2004), the protagonist undiscriminatingly plays for both sides: the USA or Nazi Germany. This political equalisation, at the very least, relates to a simple pacifist criticism which declares war to be 'another man's cause', the sending to slaughter of *Menschenmaterial* (men material). What is ignored is that there were and are actual reasons to fight fascistic regimes.

Economic catastrophe – the reign of races

Even though blunt racist games only exist as a subculture, commercially successful games such as the *Warcraft* series do not spare the 'clash of races'. Despite virtual remodelling, they suggest 'genuine' characteristics of each 'race' which are imprinted in the genetic code. What once historically served natural-justice scientists of both shades – Rousseau's 'back to nature' and Hobbes' 'homo homini lupus' – as a means to consolidate civil rights *qua* nature, in computer games is perverted to the natural necessity to contest. *Lord of the Rings*, in the game series as in the book, immortalizes good and evil, respectively, by virtue of biology. What serves as a 'crisis exit' in modern societies here is eternalized as nature's *sine qua non*. Only *Warcraft III* plays with stereotypes and allows transformations of good and evil. Mostly, however, actual racism is remodelled and universalized.

The ecological catastrophe

In computer games, natural disasters that are not made by human beings are uncommon. The 1989 version of *Sim City* included a button that the player could use to create hurricanes or set the newly built city on fire, but this was a gimmick. Ever since Chernobyl, contaminated Nature has been signified by shiny green radiation barrels and puddles to which the radioactivity symbol is affixed; but computer games deal with contaminated Nature in terms of everyday life in the virtual world. The forthcoming game *Stalker* is actually set in Chernobyl; but certain *Far Cry* (2004) missions, too, have fallen prey to contamination. The apocalyptic vision of the capitalist world's destruction via natural disasters is the topic of games such as *Fallout: A Post Nuclear Role Playing Game* (1997), *The Fall: The Last Days of Gaia* (2004), or various *Star Wars* scenarios. Even if capitalism is not spelled out explicitly, the undertone of devastation mostly refers to an unscrupulous exploitation of Nature, executed in order to maximize profits.

Genetically engineered and biotechnological catastrophe

A majority of contemporary computer games focus on a potentially apocalyptic field that is presently underrepresented in the media: abortive genetic engineering. Indeed, major class struggles in the present and future will be fought on the fields of copyright and free access to knowledge, including public domain, AIDS medication in Africa or ecological catastrophes. But at the same time a transference between genetic predominance and the prohibition of the development of rationality and free will comes about in computer games. Mutated people who serve as mere cannon fodder for the avatar's weapons are symbols of human beings as both submissive believers and failed experiments. They come out as contorted beings, as groaning creatures whose genetically conditioned bestiality results from experiments. The inflicted (natural) scientific injustice is in inverse proportion to ailing bloodlust. *Doom* (1993) in particular, deals with these unconscious fears. It is not a coincidence that *Doom's* manufacturer goes by the name Id-Software. The Freudian id describes the smorgasbord of drive and inhibitions that demands realization. The realization of the Big Fears is Id-Software's agenda, whether in *Doom* or in *Return to Castle Wolfenstein* (1983). Everywhere there are genetically manipulated creatures, driven by their instincts. The topic also appears in *Half-Life*.

Biotechnological and genetic engineering catastrophes inspire the construction of cyborgs – human beings with technological implants.

The Borg's Great Assimilation in *Star Trek* is looking for the demarcation between free will, the ability to think critically, and the engineered fate of Nature. The correlation between biotechnological and genetic engineering catastrophe consists of its victims' lack of ability to exist in terms of a humanistic conception. The player, who does not understand that these hybrids that range between human being and technology are an intrinsic part of herself in terms of modern subjectivity, considers annihilation to be the only possible consequence.

The subjective catastrophe – the failure of civil culture: psychological and religious catastrophe

Religious and psychological catastrophes are similar; they share the quest for meaning with the loss of meaning. In modern civil society, a whole industry is available to inspire meaning in the subject's existence, and love ranks first. Computer games extensively draw on this layer. Let's pretend there were modern societies without culture: they would be dreary, technological and hollow, and they would be visibly based on merely exploitative conditions.

In computer games, the catastrophe appears in the form of fallen human beings, mostly as the incarnations of evil, who are no longer open to the meaning of modern and past societies. At the same time, the amount of power they represent, and are provided with, is astonishing. They personify the degree to which culture and religion suggest it could be worthwhile to abnegate the preset way. How else can it be explained that the evil-doers are being provided with so much power, and are extensively celebrated in the games' stagings? Lucifer, the fallen angel, stands up as the prototypical godfather.

In most cases, the fallen characters represent the inversion of Kant's enlightened ruler. In computer games, they represent nothing less than the hero's negative virility and his caricature: only those who anticipate their own meaninglessness aspire to higher things, whether good or evil. This is exactly why the changing between light and dark campaigns works seamlessly in games – the player preconsciously anticipates her own marginalisation. *Lord of the Rings* or the *Gothic* game series draw on these stylistic devices. Failed and debauched beings present themselves as mighty enemies and intimidate with their threat to rule the world. Hidden behind this threat is the possibility of the player's own debasement when faced with capitalism's superiority. Conditions can only be changed by applying the same means. The player reacts, for fear of her own failing life, with that very relentnessness which the society she is afraid of threatens her with. Instead of achieving emancipated subjectivity, the one-dimensional subject falls back on supporting the brutality she is threatened with. In *Lord of the Rings* (Tolkien's novel, not the game), Sauron, the fallen elf, transcends the discrepancy between the individual's omnipotent fantasy and his actual social reality (in terms of wishing to surmount it and spread darkness and slavery upon the world). He will accordingly resemble the very image he fears, and that is inflicted on him as a personal failure. Darth Vader, too, functions according to this image.

Human beings as submissive, misguided objects, free from any subjectivity, have so far not been a differentiated topic of computer games though they appear in the form of soldiers. Human beings as war's victims exist as figures without subjective features. To kill them is all the less problematic because they emerge as threatening, remote-controlled, empty human shells with weapons in their hands.

Enter the Matrix (2003) visualizes Kant's nightmare (the human being without the categorical imperative) and Schopenhauer's fundament (The World as Will and Idea) in terms of a deeper truth hiding behind the useless organisations of existing societies that are not accessible to the individual. The concealed truth reveals itself in the assumption that the present world could be a free one – imagined from the position of developed, productive forces – but does, however, fall back on the formation of a one-dimensional consciousness. The remote-controlled being, so the matrix reveals, in reality is no less than everyday people, reflecting and reacting to an order that is neither hostile nor favourable.

To avoid misunderstandings: this article uses various scenarios that are collage-like images of our times – an observed possibility of catastrophes, as shadow realms of existence. No game is able to wholly apprehend the problems of the incipient 21st century; the problems are, however, included in those games – that is why the subject as a whole, on the one hand, involves a high degree of criticism but on the other an unbearable degree of kitsch. An example occurs in *The Sims* (2000), where even the thief's deepest fear is about her home decoration. The following archetypes of preconscious apocalypse in computer games can be summed up as follows[3]:

Economic/political catastrophe

- The game worlds present the reign of rackets (for example *Deus Ex; Vampire – The Masquerade: Bloodlines*, 2004: *Command & Conquer*, 1995) while simultaneously displaying the population's poverty
- Slums and ghettos are a common part of life (*Deus Ex; Vampire – The Masquerade: Bloodlines, Max Payne*)

- In rival gangs, subcultures fight for their survival, state governments have collapsed (*Mafia The City of Lost Heaven*, 2002; *GTA* series)
- Development of fascistic social systems (*Half-Life 2*)
- Military reign/dictatorships, occurrence of totalitarian states, war scenarios (*Codename: Panzers; Commandos: Behind Enemy Lines*, 1998)
- The battle between ethnicities/races (the *Warcraft* series; *Diablo*, 1996)

Ecological catastrophes

- Contaminated Nature as the virtual world's (daily) routine (*Far Cry; Stalker*)
- The world as an end-time scenario, an uninhabitable world (*Fallout: A Post Nuclear Role Playing Game; Star Wars* series)

Genetic engineering catastrophes

- As a result of failed experiments, mutated human beings rule the world (*Far Cry, Doom, Castle Wolfenstein*)

Technological catastrophes

- Failed amalgamations of human beings and technology, human machines with technological implants (*Deus Ex; Star Trek* series)

Psychological and religious catastrophes

- Abortive, obsolete human beings as mighty enemies who threaten the world with their dominance (*Lord of the Rings* series; *Gothic* series)
- Human beings as submissive, misled objects (*Star Wars* series; *The Matrix* series)

Apart from preconscious apocalypse, computer games do, however, display various qualities. The heroic avatar shows signs of obstinateness, evil's seductions are ineffective, and sometimes there seems to be hope that the world may in the end be a good place.

In some quarters, beauty can be found – hope, a walk through the dreamworld, the miracle, magic, the happy ending – experiences missed in reality rest in the game. A walk through *Gothic's* dreamworlds, or cruising through *GTA San Andreas* or *Need for Speed: Most Wanted* (2005) sometimes is more fun than the next mission. For a short while, the player may forget civic morality, which is that murder is forbidden but breeding is permitted, while sexuality may not be depicted and murder may be shown on screens.

Michel Foucault once wrote that discourses about the sexual spark off everywhere, but in the end are concealed. The same is true in computer games. Everywhere discourses about violence are present, though they are ultimately oblique; and eventually computer games are nothing but games. Nevertheless, the cruel prehistory that ruling classes have bequeathed to their subjects is familiar: bread and circuses are not an invention of the modern age. All along, there has been a change between enemy, avatar and player – the game within the game consists of the reduction of reality to gaming. Brutality is no attainment of our own time, nor are conservatives' lamentations.

Whenever conservative newspapers declare computer games to be the root of all evil, one has to consider that both newspapers and computer games are articulations of the same capitalistic reality. Accordingly, the same criticism conservative newspapers level at computer games lies in these censors' claims: that they are a reduced, dreamless and one-dimensional subject, a lamentation without utopia which strives to make 'today' seem to be the best of all possible worlds.

Computer games are games, and if a society does no longer trust its subjects to realize the difference, then this tells us more about society than about its games. Young Marx said, 'Every country has the press it deserves'. The same unconditionally applies to computer games.

Notes

1. Lumpenproletariat – Encyclopedia of Marxism (n.d.) *MIA: Encyclopedia of Marxism: Glossary of Terms*, http://www.marxists.org/glossary/terms/l/u.htm# lumpenproletariat [Accessed 4 April 2006].
2. Totalisation – Encyclopedia of Marxism (n.d.) MIA: Encyclopedia of Marxism: *Glossary of Terms*, http://www.marxists.org/glossary/terms/t/o.htm [Accessed 4 April 2006].
3. The list of games is as incomplete as a bibliography of the human psyche would be. Unfortunately the author hasn't got enough time to extensively play computer games.

14
Borders and Bodies in *City of Heroes*: (Re)imaging American Identity Post 9/11

Nowell Marshall

In *Virtualities*, Margaret Morse (1998, p. 126) problematizes the idea of the cyborg posited by those in 'future-oriented subcultures who have wholeheartedly embraced technology' because 'the actual status of the cyborg is murky as to whether it is a metaphor, a dreamlike fantasy and/or a literal being'. In particular, Morse (1998, p. 125) cites the problem of inhabiting both an organic and a virtual body: 'Travelers on the virtual highways of an information society have at least one body too many – the one now largely sedentary carbon-based body resting at the control console that suffers hunger, corpulence, illness, old age and ultimately death'. In arguing for the irrelevance of what she terms the organic or 'meat' body, which she asserts 'just gets in the way', Morse privileges the virtual body. However, given recent advances in gaming technology, such as the Massively Multiplayer Online Role-Playing Game, such claims about the virtual body merit reconsideration. Although text-based computer games such as *Adventure* and *Zork* and pen and paper role-playing games such as *Dungeons and Dragons* date back to the 1970s, *Ultima Online*, which popularized the genre in the late 1990s, is generally considered the first modern MMORPG. With subscribers numbering from hundreds of thousands to millions, these online communities provide a unique opportunity to explore the bodily discourses circulating in online environments.

The worlds that MMORPG players inhabit constitute a *virtual space*, what Allucquère Rosanne Stone (2000, p. 506) terms 'an imaginary locus of interaction created by communal agreement'. In the gaming community, the virtual worlds created within this virtual space are called *persistent worlds* because they continue to exist and change even when players are absent. Upon logging in, players interact with the persistent world by using an avatar – a graphical representation of the character they play.

Through this avatar, Morse (1998, p. 17) argues, 'the user electronically wraps him- or herself in symbols by means of electronic clothing'. Although Morse originally used the term 'electronic clothing' when describing earlier virtual-reality systems that required the user to wear a head-mounted display or data gloves for tracking hand gestures, the term aptly describes virtual spaces where users customize the appearance of an avatar, thereby electronically clothing a virtual body in customizable pixels of light. However, some aspects of the online immersive process that allow players to customize the body of their avatars become problematic. If, as Arthur Asa Berger (2002, p. 5) argues, video games function as indicators of culture 'similar to the novel', then constraints on avatar customization within MMORPGs mirror contemporary discourses of the body. One of the most popular of recent American MMORPGs, *City of Heroes* (2004) offers problematic representations of the body that reveal the game's articulation of a post-9/11 American identity privileging conflict and the maintenance of rigid, xenophobic borders against a variety of aberrant bodies.

Players new to *City of Heroes* have the option of completing a tutorial, which teaches them how to manipulate the game's interface to use powers and enhancements and select missions. By successfully completing missions, the player earns experience and advances through what the narrative labels the hero's security level. Both criminals and heroes have security levels with a hero's chance to hit an enemy being determined by the difference in levels between the character and the enemy. Paragon City, the game's persistent world, is divided into city zones by enormous energy walls. From a technical perspective these impenetrable, blue walls exist to mitigate bandwidth problems and shorten loading times when travelling through the virtual landscape. However, the game's narrative specifically posits these walls as protective systems, which divide the city into safe and unsafe zones by prohibiting enemy movement from zone to zone and allowing players to move between zones by using tunnels and light-rail systems operated by the Paragon Transit Authority. Beyond various city zones and the city sewer system, players who have obtained the appropriate security clearance can also enter trial zones, such as Perez Park, the Hollows, Dark Astoria and Crey's Folly. Although some critics, such as Morse, champion such visible boundaries because they reveal the constructed nature of virtual space, the maintenance of borders, which figures prominently throughout *City of Heroes*, does not always amount to progressive representation. In addition to the visible barriers that segment the virtual landscape, the game features a variety of criminal organizations and gangs, each with unique goals, backgrounds and superpowers. The backgrounds of these various organizations, which players uncover by completing missions, and in many cases the visual construction of their criminal bodies, reinforce the delineation of borders and bodies within the game. In effect, bodies become borders distinguishing good from evil, hero from criminal, and ultimately the productive American from the destructive foreigner, the alien and the monstrous queer.

When creating a character in *City of Heroes*, users makes several choices, exercising what Janet Murray (1997, p. 126) terms *agency*, 'the satisfying power to take meaningful action and see the results of our decisions and choices'. Yet, the game interface limits users' agency during character creation and later during game-play itself because, as Murray (1997, p. 129) argues, 'when we move narrative to the computer, we move it to a realm already shaped by the structures of games'. Thus, the pre-existing structure of the game coded by software engineers inscribes and limits player agency. The choice of body types available to players in *City of Heroes* provides an example of this limited form of agency, refuting Sandra Calvert's utopian claim (2002, p. 58) that 'virtual bodies, made possible on the Internet' allow people to create 'any kind of body that they want to present to others. The fat can be thin, the short can be tall, the weak can be strong'. While Calvert's argument holds in some contexts, her argument is too broad to sustain in the context of MMORPGs. First, the desires of internet users, whether enacted through a MMORPG or a simple chat-room interface, always encounter limits set by the designers of the virtual space and by the interface itself. Furthermore, Calvert's argument privileges normative social ideals. She focuses on a 'fat' person's ability to perform 'thin' in a virtual space; however, very few virtual spaces allow users to engage culturally undesirable roles. Thus, while a MMORPG might allow a 'fat' person to perform 'thin', chances are much less likely that someone 'thin' will have the option to perform 'fat'. Few MMORPGs allow for such non-normative options in avatar design.

In *City of Heroes*, players must select an origin; however, unlike the racialized origins available in most fantasy MMORPGs, which enforce a system of racial benefits and weaknesses, origins in *City of Heroes* determine only the player's initial contacts and usable types of enhancements. Available origins include natural, where abilities are learned through training (e.g., the military); mutation, which focuses on genetic evolution (e.g., Marvel's *X-Men*); science, where scientific experiments yield powers (e.g., Marvel's *Spiderman*); technology, where devices enable powers (e.g., Marvel's *Ironman*, DC's *Batman*); and magic (e.g., Marvel's *Dr. Strange*). Although aliens comprise a significant number of heroes in traditional comic-based mythology (e.g., DC's *Superman*), they remain unavailable to players in *City of Heroes*. This omission of aliens as an available character origin plays an important role in the narrative structure of *City of Heroes*, which accentuates the conflict between human and alien to construct a specifically post-9/11 American identity through the conflation of bodies and/as borders.

In considering how *City of Heroes* constructs virtual bodies as borders, Eugene Thacker's (2004, p. 12) distinction between *the body* and *embodiment* proves useful:

> Whereas 'the body' relates to those social, cultural, scientific and political codings of the body as an object (public/private, racialized, gendered, anatomical), 'embodiment', as Maurice Merleau-Ponty relates, is the interfacing of the subject as an embodied subject with the world.

Given Thacker's distinction, restricting discussion of virtual bodies, or avatars, to what Thacker terms *the body* and what Judith Butler (1990, p. 12) describes as 'a passive medium on which cultural meanings are inscribed' proves beneficial. This distinction becomes particularly fruitful when considering theorizations of the body by Stone, who discusses both *the body* and *embodiment*. Of avatars, Stone (2000, p. 518) writes, '[b]ecause cyberspace worlds can be inhabited by communities, in the process of articulating a cyberspace system, engineers must model cognition and community; and because communities are inhabited by bodies, they must model bodies as well'.

At this point, Butler's (1990, p. 13) question, '[t]o what extent does the body *come into being* in and through the mark(s) of gender?' becomes crucial in analysing available bodies in *City of Heroes*, which initially restricted avatars to nearly idealized bodies. The game interface allowed players to design characters ranging from roughly four to seven feet tall within a limited weight spectrum ranging from athletic to muscular. On the surface, this restriction adheres to comic-book conventions; however, mainstream comics such as Marvel's *Generation X* (1994–2001) featured 'misshapen' and 'disfigured' heroes (e.g., Skin, Chamber). To compensate for this deficiency, game designers expanded avatar customization options in May 2005, allowing players to either customize their avatar's physique, shoulders, chest, waist, hips and legs or select pregenerated bodies labelled Slim, Average, Athletic and Heavy. Although such concessions in bodily construction indicate progress in the game's depiction of bodies, several other areas within the game remain problematic.

Although Tim Gill (1996, p. 39) argues that '[c]omputer games undeniably draw on stereotyped images of gender', critics such as Stone have argued for a more utopian, or at least progressive, notion of virtual space. Stone (2000, p. 520) argues that while software engineers 'are surely engaged in saving the project of later twentieth-century capitalism, they are also inverting and disrupting its consequences for the body as object of power relationships'. As a test case, *City of Heroes* both affirms and ruptures the utopian potential that Stone sees in virtual engineering. The game affirms Stone's argument by offering an escape from the traditional male/female binary by offering a third body type, which the game terms *huge*. This enormous body type is necessary to recreate many well-known comic-book characters (e.g., Marvel's Beast, Juggernaut). However, the theoretical value of this third type becomes complicated. In *City of Heroes*, the character creation screen asks the player to 'Choose a Body Type'; yet, the body types themselves are labelled in gendered terms: female, male, huge. In this conflation of gender and body type, the huge body, which clearly *looks* male and offers customization features similar to those of the male avatar, is simultaneously coded as both male and not male. Within this three-gender system, the huge body suggests a monstrous body, creating a space of otherness within the game. Indeed, all characters based on the huge body template move in a hunched manner that further correlates the huge, male-yet-not-male body with monstrosity.

From a Western European cultural perspective, the concept of monstrosity has a long history of interpretation as a transgressive or queer masculinity dating back to the 17th century. Michel Foucault (1999) aligns the monster with social and sexual deviance. Likewise, reviewing the rhetoric employed in Mervin Touchet's 1631 sodomy trial, Gregory Bredbeck (1991, pp. 5–6) argues that sodomy and homoeroticism were 'contained within a mythology of the unnatural, the alien, and the demonic [...] attributing sodomy to *foreign* languages and *monstrous* men'. More recently, Judith Halberstam (1995, p. 3) argues that 'the emergence of the monster within Gothic fiction marks a peculiarly modern emphasis upon the horror of particular kinds of bodies'. Although *City of Heroes* may not seem like a Gothic narrative, the game promotes the fear of invasion and the desire to protect, tactics that Halberstam (1995, p. 2) specifically notes in defining the Gothic as 'the rhetorical style and narrative structure designed to produce fear and desire within the reader'. Following Morse's argument (1998, p. 185) that '[v]irtual landscapes can also figure as liminal realms of transformation, outside of the world of social limits and constraints', reading the huge body as a monstrous or queer text opens possibilities for progressive politics within in the game's otherwise reactionary narrative. Yet, because the huge body type itself remains inscribed within the discourses of masculinity, it offers only marginal potential for the type of 'radical proliferation of gender, to displace the very gender norms that enable the repetition itself' that Butler (1990, p. 189) envisions. In *Remediation*, Jay David Bolter and Richard Grusin (1999, p. 238) argue that '[i]n its character as a medium, the body both remediates and is remediated. The contemporary, technologically constructed body recalls and rivals earlier cultural versions of the body as a medium'.

Thus, in offering the huge body type, *City of Heroes* rearticulates the male/female binary as a slightly expanded three-gender spectrum that at first seems promising. However, this delineation of the huge body as both male and not male also resurrects early psychoanalytic stereotypes of the butch homosexual, what Sandor Ferenczi (1952, p. 300) in his 1914 paper 'The Nosology of Male Homosexuality' termed the 'object homo-erotic':

> He feels himself a man in every respect, is as a rule very energetic and active, and there is nothing effeminate to be discovered in his bodily or mental organization. The object of his inclination alone is unchanged, so that one might call him a homo-erotic through exchange of the love-object, or, more shortly, an object homo-erotic.

Ferenczi argues that although the object homo-erotic identifies as male, he is conscious that his desires are not traditionally masculine. As a result, the huge body type in *City of Heroes* may represent the resurrection of early 20th-century classifications of the object homo-erotic depicted in games as the male body that is both male and not male with excessive muscular development functioning as a type of 'compensation'. This reading of the huge body mitigates the utopian potential that Stone attributes to virtual space because by reviving early 20th-century stereotypes, the game remediates the monstrous body within a reactionary framework.

Although the game diverges (albeit problematically) from the typical male/female gender binary, it also enforces normative models of beauty derived from 18th-century aesthetics. In his discussion of the sublime and the beautiful, Edmund Burke (1998, p. 148) distinguishes smoothness as 'a quality so essential to beauty, that I do not now recollect any thing beautiful that is not smooth'. *City of Heroes* posits smooth, neoclassical bodies, like the statue modelled on the Greek god Atlas in Atlas Park, not only as normative and heroic, but also as standards of beauty. These enormous, neoclassically smooth monuments appear in all of the game's major zones providing a role model to players, who must maintain the social order of Paragon City by repelling foreign bodies that do not conform to such neoclassical beauty standards. Such neoclassical conceptions of beauty become instrumental in how *City of Heroes* distinguishes between heroic bodies and criminal bodies: whereas heroic bodies conform to neoclassical standards of beauty, remaining 'free from all foreign admixture,' criminal bodies violate this edict (Winckelmann, 1996, p. 129). Given Halberstam's articulation (1995, p. 6) of the monstrous body as that which 'makes strange the categories of beauty, humanity and identity', the majority of enemies in *City of Heroes* read as monsters.

Although players encounter many types of criminals, the game's narrative revolves around the Rikti invasion on 23 May 2002:

> The war waged on for the next six months, during which time hundreds of thousands of soldiers, civilians and heroes died in battle. The Rikti proved a decidedly intractable foe. They had apparently been coming to Earth well before the actual invasion began in earnest, setting up hidden bases and weapons caches beneath the ground. They used short-range teleportation portals to strike at unexpected times and locations (*City of Heroes* – Game Info: Geography, n.d.).

This narrative emphasis on American deaths, invasion, secret plots and unpredictability mirrors the aftermath of 11 September 2001 terrorist attacks on New York City. The splash screen that loads when *City of Heroes* launches illustrates this tie to 11 September. The central hero of the image, the Statesman, wears a red, white and blue costume complete with a large white star on his chest, invoking the United States flag. The skyline on this splash screen resembles a pre-9/11 New York skyline. The game's website describes Paragon City as 'the greatest city in America', setting the game not only in America, but in America's paragon of cities, New York. Thus, the 2001 terrorist attack on New York plays a significant role in the game's narrative structure.

The game depicts the alien Rikti as the US media has characterized terrorists in the aftermath of 9/11. Right clicking on a Rikti Communications Officer reveals the following information: 'The Rikti have come to stand for everything humans hate and fear in the universe. Their unprovoked attack invasion of Earth has left much of the planet traumatized and angry.'

The Rikti (reminiscent of H.R. Giger's creatures) as well as other enemies within the game, including the Banished Pantheon, the Devouring Earth, the Lost, and the Freakshow, possess asymmetrical, imperfect and unsmooth skin, confirming Halberstam's suggestion (1995, pp. 6–7) that skin 'becomes a kind of metonym for the human; and its color, its pallor, its shape mean everything within a semiotic of monstrosity'. Furthermore, many of the special events introduced into the game revolve around the threat of invasion. One such special event titled 'Heroes all over the city band together to fight off potential invasion force' launched on 17 September 2004. The event, announced through the *Paragon Times* on the game's website, introduced purple portals through which extradimensional aliens attacked Paragon City. These aliens are particularly relevant for their deviance from the game's normative beauty standards. Instead of appearing as symmetrical, smooth bodies, the majority of the aliens featured large, club-like limbs and rough skin. To heighten suspense, special-event aliens were randomly generated within city zones, leaving players to wonder when and where their enemies will next strike and providing yet another point of comparison between the game's preoccupation with alien invasion and American fears of terrorist attack after 9/11.

Another *Paragon Times* article titled 'A New Type of Alien Visitor' related the account of a reporter contacted by the alien Kheldians, 'beings composed only of energy. When they came to Earth, they found they were able to merge with humans, which can benefit both parties' (*City of Heroes* – Newspaper, n.d.). However, introducing the Kheldians quickly became another method of reinforcing the conflict between human and monstrous inhuman, American and foreign invader within the game because the benevolent Peacebringers 'are not the only Kheldians on Earth. There is another group, called the Nictus, who are not concerned with the needs of those they merge with'. As expected, the *Paragon Times* article concludes by referencing the ubiquitous threat of the monstrous alien, reinforcing the paranoid rhetoric of post-9/11 America, and praising military buildup: 'As we learned painfully from the Rikti Invasion, we are not alone. We cannot bury our heads and hope that we will remain safe. The silver lining to this grim warning is that we have more heroes than ever dedicated to protecting this city against whatever threats may come' (*City of Heroes* – Newspaper, n.d.).

Morse (1998, p. 3) agues that '[t]he allure of television has deep roots in the need for human contact and the maintenance of identity and for a sense of belonging to a shared culture'. Given the similarities between television screens and computer monitors, her argument extends to MMORPGs, which offer the promise of community through trade and combat guilds. Yet, in an age when technology offers vast potential for designing virtual spaces, why have software engineers chosen to create such dystopian national narratives?

In depicting the Rikti, and by extension aliens in general, as standing 'for everything humans hate and fear in the universe', *City of Heroes* departs from the comic-book tradition that inspired the game. Whereas the alien and the monstrous in science fiction and fantasy computer games often pose a threat to human society, these figures have a broader history in comics. Although some comic-book aliens, (e.g., Marvel's *Galactus*, DC's *Sinestro*) routinely threaten human society, many others (e.g., Marvel's *Shi'ar Empire*, DC's *Superman*) have become invaluable allies of humans. While many aliens in *City of Heroes* morphologically resemble aliens in the Marvel and DC universes, the narrative purposely deviates from the more balanced comic-book mythology to provide the alien as a post-9/11 scapegoat, a foreign monster functioning as its focal point and fostering a dystopian virtual narrative that advocates symbolic violence within Paragon City.

Murray (1997, p. 129) emphasizes 'spatial navigation', arguing that even dystopian environments can provide pleasure 'independent of the content of the spaces'. Yet, billboards built into the scenery of Paragon City proffering rhetorics of paranoia and defence mitigate this utopian sense of agency that Morse identifies. One billboard has a navy blue background with white lettering that reads 'THEY COULD BE ANYONE' across the top left and continues, 'Be careful...' along the bottom left corner. This common billboard, which appears near many light-rail stations and enhancement shops, instils a sense of paranoia in players. Another sign features the silhouette of a city skyline complete with dual towers. Hovering above the skyline against a red background suggesting threat, a pair of eyes glowers at the player. Stamped in bold, black text outlined in red across the top of the sign are the words, 'THEY ARE STILL AMONG US...' followed by white lettering along the lower left of the sign that reads, 'Report Any Strange Activity to the Vanguard'. A partial American flag that fades into a portrait of the Statue of Liberty comprises a third sign's background. Across the top of the sign in white letters, reads, 'EARTH FOR HUMANS' and along the lower left, 'Let's Keep It That Way!' Although the text of this sign proclaims Earth for humans, the American icons comprising the background clearly convey a different message, reducing 'Earth' to the United States and 'humans' to United States citizens.

Furthermore, the backgrounds of the criminal groups in *City of Heroes* become problematic. Many groups, such as the Circle of Thorns and the Banished Pantheon, are criminalized specifically because they adhere to non-Christian religious systems. Depicting non-Christian religious practices as a threat furthers the game's post-9/11 nationalist rhetoric, paralleling the increasingly strained relationship between Americans and Muslims in post-9/11 America. At times, the game narrative also evinces virulent homophobia. For example, one mission asks players to terminate Patient Zero, an infectious member of the Vahzilok. Leo Bersani (1987, p. 202) traces the name Patient Zero to Gaetan Dugas, 'the French Canadian airline steward ... responsible for 40 of the first 200 cases of AIDS reported in the US'. Thus, this mission asks players to save the *general population*, which Jan Zita Grover (1987, p. 23) argues always means *the heterosexual*, from the threat of HIV infection. Likewise, missions involving the Devouring Earth (environmental activists depicted as mutated plants) require players to rescue hostages. When rescued, these hostages say, 'They were going to change us!' or 'They were going to change the children!' Numerous missions in *City of Heroes* posit change as a threat to futurity through the physical and ideological conversion of children or as a threat to the capitalist social order through the conversion or sacrificing of productive members of society. By explicitly opposing foreign/alien, non-Christian, and monstrous/queer bodies to heroic, neoclassical, American bodies, the narrative inscribes social, religious, national and sexual difference on to criminal bodies and then directs players to enact violence against these bodies. If, as Murray (1997, p. 183) contends, 'cyberspace has the potential to be the most powerful and effective means of surveillance and social control, not only of the user in cyberspace, but of the external material world', then the social impact of bodily discourses circulating within persistent worlds such as *City of Heroes* merits further investigation.

15
Anti-PC Games: Exploring Articulations of the Politically Incorrect in *GTA San Andreas*

Andreas Jahn-Sudmann and Ralf Stockmann

'Ultimately, if you are over 18, love videogames, and heaping doses of political incorrectness and foul language do not bother you, I would have a difficult time believing that you would not enjoy this game.'

JasonEthos, Reader Review (2004)

To date controversies about computer games primarily focus on the representation and simulation of violence and their presumed harmful effects, particularly on children and adolescents. This connection of violence and computer games and its damaging potential is deeply anchored in the public's mind. However, debates on the presentation of sexuality in computer games, normally a traditional focal point of the possible socially damaging consequences of media consumption, are rare. This may initially be ascribed to the fact that the representation of visible, explicit sexuality is rather unusual in computer games due to their assumed status as a children's and youth-entertainment medium. In our case study of the both popular and controversial game *GTA San Andreas* (2004), the articulation of explicit sexuality and violence in computer games unquestionably plays a decisive role.

As a contribution to critical game studies, however, an aspect of representation should be emphasized that neither confines itself to the display of sex or violence and yet is relevant to the intrinsically controversial content of the game: the articulation of political incorrectness.

PC/Political incorrectness: alleged censorship/pseudo taboo breach

Since the beginning of the 1990s, political correctness (PC) has been used as an umbrella term referring to an alleged practice of censoring language and opinions that reputedly represses truths, violates fundamental democratic rights by installing an absurd language control, and abrogates traditional values and cultural traditions. Correspondingly, a community of PC-adherents, as well as a putative consistency of what defines PC, is constructed around content. Among others, Marc F. Erdl (2004) has criticized this position in his profound study, and depicted the public discussion about political correctness as a mock debate. There have been and are no exponents of a political correctness and accordingly, no political programs or manifestos that could be subsumed under the term PC (Manske, 2002, p. 28).[1]

This knowledge has not prevailed, resulting in the consistent application of the term PC in a pejorative manner. To the extent that PC has been established as a negative term, political incorrectness could be promoted as a positive, widespread concept of hegemonic culture. Political incorrectness became the epitome of legitimate provocation, of identity formation based on nonconformity, and of the subversion of the 'PC-subversion'.

Apart from this rather commonplace meaning, three categories of the usage of ideas such as political incorrectness can be identified, including political(-ly) incorrect(-ness):

1. As a calculated, taboo-breaching, provocative (culture-industrial) sales approach. This refers to the attraction value of the politically incorrect in terms of satire, jokes, art or as a realistic (read: 'permissive') description of a condition. This context also functions as a framework of legitimation. Significantly, this category cannot be assigned to a simple Left/Right or Progressive/Conservative scheme. Accordingly, the popular ABC-TV talk show, *Politically Incorrect*, which describes itself as 'one of television's most provocative and funny shows', can be included in this category.

2. As a conservative counter-discourse, as an explicit counter-term and practice that discredits the imaginary notion/entity of PC. In the course of the PC debate, a vast number of (text) books have been published that seek to highlight the PC problem and its reputed potential for intimidation (for example Elder, 2000; Zilbergeld, 2004).

3. As an explicit and intended discriminating representation and propaganda practice that obviously is not (only) concerned with 'humour' but with de facto existing racist, anti-Semitic and sexist dispositions that are articulated openly and in public. This category rarely turns up in direct association with the term politically incorrect. But it does exist. One example is the racist and anti-Semitic computer game *Ethnic Cleansing* (2002) promoted and distributed by the label Resistance Records which belongs to the Neo-Nazi organisation National Alliance.

Political incorrectness as a concept of analysis?

In the following study of *GTA San Andreas,* we presuppose that analysis of politically incorrect computer games does not only involve those games that either by self- and/or extrinsic attribution have been described as politically incorrect. In fact, it is imperative to take a look at the very cultural articulations and their contexts in which labels like 'anti-PC' or 'politically incorrect' do not appear explicitly, but rather emerge as politically incorrect articulations. This inevitably leads to considerations of whether (theoretical) assumptions and ideas of political incorrectness can be transferred to critical analysis at all. On the one hand, the presumption of the 'hermeneutics of suspicion' or 'paranoid readings' suggests that, in the course of a critical analysis, certain cultural artefacts are being classified or exposed as politically incorrect. Self-proclaimed PC-opponents would interpret this as just another proof of their thesis that such a critical analysis has nothing but censorship in mind. On the other hand, one would have to dispose of a (heuristic) definition of the politically incorrect that produces problems such as hasty assessments and normative predefinitions.

Nonetheless, it is useful to establish a pragmatic minimal definition of the term politically incorrect. According to our understanding, politically incorrect representations affect sociocultural formations with a subordinate status (homosexuals, women, ethnic minorities etc.) that are displayed in a negative or discriminating way, whether by negative stereotyping, omission or in any other way, in terms of a calculated provocation or a wilful breach of taboo. In this context, the intention not to be PC, articulated in the game and consequently recognizable, is definitely more important than the actual content of the politically incorrect articulations. The recipient is expected to know that in the actual product, textual strategies that undermine the politically correct exist. The debasement of social minorities, however, is the core of politically incorrect articulation as we understand it.

GTA San Andreas – game plot

GTA San Andreas (2004) is the fifth sequel of the extremely popular and controversial *Grand Theft Auto* series originally developed by the Scottish game designer Dave Jones (DMA-Design, later Rockstar North). The game is based on an alternation of car (racing) sequences and third-person-shooter insertions. Adapted from the role-playing genre is the combination of level missions and game zones that can be configured independently so that the player can pursue her own interests, such as taking up jobs or cruising the game world. The 'arming' of the avatar and its motor pool, respectively, also follow role-playing conventions. The tuning of automobiles conforms to cultural images of an ethnically construed macho culture (for example, in MTV's *Pimp My Ride*) and is *mutatis mutandis* transferred to the avatar. Ranging from hairstyle, apparel and tattoos to the build-up of muscle mass, the character can be shaped within given boundaries.

The actual plot of *GTA* describes the rise of a young African-American, Carl 'CJ' Johnson, from petty criminal to gang leader in a fictitious state on the West Coast of the US in the early 1990s. The urban game world is plagued by delinquency and violence, beginning with minor damage to property and larceny and culminating in murder. A recurring theme is the family's or gang's 'reputation'. The segregation of the avatar's group from other groups is primarily based upon ethnic affiliations. The respective gangs and their milieus are clearly encoded by their ethnicity as African-Americans, Latinos/Latinas, Italian-Americans or Asians. New to the *GTA* series is the integration of 'heterosexual relationships'. CJ has the opportunity to 'win' up to six girlfriends. Even though these relationships are not the centre of the game's missions, they are revealing as related to the specific articulation of the connection of race and sex/gender.

The representation of African-Americans in computer games

It took decades for African-Americans to attain the degree of public visibility and relative prominence in the media that seems so natural nowadays. But this development did not occur simultaneously in every popular cultural medium. In the world of electronic gaming, African-Americans – and non-whites generally – have rarely been featured as playable central avatars. Their representation is primarily restricted to the field of sports or beat-'em-up games (Leonard, 2006, pp. 83–8). Outside these genres, one of the first black avatars can be found at the intersection of role-playing and adventure games: Michael F. Stoppe in *Maniac Mansion* (1987). It is noteworthy, however, that Michael is not featured on the game's packaging, which depicts only five out of the seven selectable avatars. Furthermore, an African-American character as a given (and not removable!) avatar can be found in the action game *Urban Chaos* (2000), where the protagonist is a female Jamaican-American police officer named Darci Stern.

The question is, does the depiction of black avatars follow the PC-politics of representation of contemporary Hollywood movies? On the surface this depiction is contingent on the identifiable efforts to avoid a discriminatory representation of African-Americans (as well as other minorities). By now it is standard in contemporary Hollywood practices that African-Americans not only portray heroes but entirely 'ordinary' citizens without any noteworthy differences from Caucasian culture.

The portrayal of *GTA San Andreas*'s protagonist, CJ, presents a completely different picture. He is the typical personification of black gangsta hip-hop culture which in the early 1990s became enormously popular. That popularity was reinforced and developed in the cultural artefacts of the global media industry, beginning with gangsta rapper video clips by Ice-T and the Notorious BIG to the independent and studio films of the New Black Cinema such as *Boyz N the Hood* (1991), *New Jack City* (1991) or *Menace II Society* (1993). These cultural articulations follow a logic of representation that is diametrically opposed to Hollywood's integrative visions.

Cultural geography: 'Welcome to the black gangsta world!'

GTA San Andreas is set explicitly in the urban world of 1990s ghetto culture and does in many respects refer intertextually to the popular cultural texts of this period. It is, however, not the case that one popular cultural product merely refers to another. At the end of the game, various missions expressly take up the 1992 LA riots. Following the actual events, a police scandal in the early 1990s leads to riots in the 'Los Santos' (read: Los Angeles) area. *San Andreas*'s game world appears realistic insofar as references to the 'real' world are clearly recognizable.

The game world's three big cities – Los Santos, San Fierro and Las Venturas – are distinctly influenced by the actual cities of Los Angeles, San Francisco and Las Vegas. Various buildings, hotels, bridges, streets or districts can be located in San Andreas, though under different names: Hollywood is now called 'Vinewood', 'Mullholland' in Los Santos is Beverly Hills, and The Luxor in Las Vegas turns up as 'Carlos Del Aztecas' and so on. But despite those references, San Andreas is not a world that seriously invites verification of its credibility in terms of an empirical or psychological realism. It rather is a bizarre universe of felony ruled by gang wars, organized crime, shootings and drugs.

The one-dimensional black man

It is a basic characteristic of avatars that they are shaped only in part by the game designers. In fact, the avatar's character is also stamped by the player and her actions with the freedom of action in turn laid down in the programmer's code. Within the scope of this article, we are primarily interested in the avatars' predefined attributes that are independent of the player's decisions and intentions. These are not only CJ's complexion, sex and sexual orientation but also his identity as a violent criminal. The combination of the identity construction of black/male/heterosexual/ violent/delinquent in many respects has a long tradition in popular cultural representations, especially in those of film. Think of D.W. Griffith's silent movie classic *Birth of a Nation* (1915) and, though hardly comparable, the so-called Blaxploitation films of the 1970s or New Black Cinema. Henry Jenkins (cited by Walker, 2003) referred to the third sequel of *GTA* as computer games' *Birth of a Nation* and thus lifted it to the rank of a culturally important masterpiece. Interesting in our context, and also considered as such by Jenkins, is whether *GTA San Andreas*, from a politics-of-representation point of view, carries dubious implications that are similar to Griffith's film, especially in the depiction of 'race'.

The black gangsta is a character who, inside and outside of black culture, radically runs counter to the claims for positive role models as they are superficially satisfied in Hollywood movies such as *Philadelphia* (1993) or *Independence Day* (1996) (See Jahn-Sudmann, 2006, pp. 150–6). CJ is no exception to this rule. In addition, the cosmos of criminals through which you move your character is almost exclusively marked by ethnicity. There are hardly any white criminals.[2] In the films of the New Black Cinema, white violence, such as that of policemen, is depicted as a counterpoint and context for the anger and violence of black youths who feel unfairly treated. Even when a black policeman takes brutal action against black youths this can be read as an expression of self-hatred, which illustrates the pressure to conform to an assumed white culture (but also as the not unproblematic imagination of the denial of a *specific* cultural identity). In *GTA San Andreas* there are no such contextualisations of violence; they would not fit into the gesture of the politically incorrect. Quite the reverse, they could be understood as the expression of a PC-conformable discourse.

Style, coolness, violence and irony

A central characteristic of *GTA San Andreas*'s game worlds, as well as of its protagonist CJ, is the connection of style, coolness, irony and excessive violence which are also characteristic of the diegetic cosmos and protagonists in Quentin Tarantino's *Pulp Fiction* (1994) (See Willis 1997, pp. 189–216). As in *Pulp Fiction,* cultural codings of black ghetto culture in *GTA San Andreas* primarily function as cultural artefacts, as a repertoire of style, isolated from any historical or social embedding. In Tarantino's film the articulations of black ghetto culture refer to other popular cultural artefacts of media culture rather than to actual events or discourses. Any political or social trace of reference to the enunciative conditions of those articulations of a culture that is depicted/depicts itself as African-American is extinguished (Willis, 1997, p. 212).

Stripped of these references, black culture functions as a cultural repertoire for imaginary appropriations, not least those of a white heterosexual (male) culture. Game and film allow white heterosexual middle-class subjects to take the place of *the other* without leaving their own place. Cultural insiders can imagine themselves as outsiders and take a comfortable trip into *difference*. Preceding this act of imaginary appropriation is the neutralization of the social and political category of race. At the same time, typical stereotyped constructions of race which are an integral part of hegemonic culture are preserved, especially the equation of blackness with style, coolness and violence/delinquency which, as opposed to the 1970s Blaxploitation films, no longer shows any political dimension. On the contrary, the semantic junction is framed ironically, refracted and expanded. The connection of style, coolness and violence/delinquency related to an ironic, humorous and comic-like embedding is decisive for our analysis of the politically incorrect. On the one hand, it constitutes the ostentatious gesture of the alleged subversion; on the other hand, it provides the legitimizing framework that allows both one-sided negative stereotyping representation of African-Americans and particularly sexist inscriptions.

Beyond 'Hot Coffee' – race, sex and violence in *GTA San Andreas*

Generally, age ratings of computer games, especially in the US, react more sensitively to the depiction of sexuality than violence. This phenomenon can also be observed in the rating system for Hollywood movies. *GTA San Andreas* illustrates this disparity graphically. Successfully completing the game necessitates serial murders. By means of cheats it is also possible to activate hidden contents and abilities – implemented by the programmers and thus obviously there to be discovered – to increase the body count almost arbitrarily.

All along, violence in computer games has been offered as a strategy to resolve conflicts. As early as 1976, the very first game, *Death Race*, was removed from the market because of its violence. Officially, the game's aim was to run down 'graemlins' and gather points within a given time frame. Due to the very reduced and abstract graphics these 'graemlins' could, however, well be perceived as humans.

The body count, the number of enemies killed or to be killed, was already considerably high in early games. An escalation thus is bound to an increase of realism (more detailed graphics and better sound) or to a specific embedding of violence. In the 1980s, the latter, not restricted by technology, produced a range of obviously politically incorrect games. *Commando Lybia* (1986), for example, included executions and pronounced itself 'Sadism Game of the Year'.[3]

A particular form of politically incorrect games, especially in terms of the representation of race, are so-called hate games.[4] The titles alone of games such as *Ethnic Cleansing, Shoot the Black, White Power Doom* or *Concentration Camp Manager Millennium* do not offer a broad range of interpretations with regard to the attitudes and intentions that underlie them. Their objective is to actively exterminate the same groups of opponents: African-Americans, Arabs and Jews.

But beyond racist and anti-Semitic hate games, in many other computer games race plays a decisive role in the construction of the enemy. Many first-person shooters employ fantastic creatures as enemies that can easily be identified as 'alien races'. Frequently animal races, such as wild cats, insects or reptiles, act as a blueprint. Although these enemies are clearly provided with consciousness and in part superior intelligence, killing them is legitimated by the construction of their racial difference. Race as an uncontested category of difference also configures the role-playing genre: the choice of the heroine's (fantasy-) race – human being, dwarf, elf, etc. – defines her future path of life from the start and can only be changed within appointed limits.

In *GTA San Andreas*'s race discourse, however, new boundaries emerge: the enemies are not defined by means of fantastic races but by 'real' racial constructions. Completely apart from the release of the so-called 'Hot Coffee' patch, there was no debate about the dubiousness of available cheats that enabled the player to run amok in the simulated world. Similar to a cheat, the Hot Coffee patch activates hidden features of the game. In this case, the typically obscured sexual interactions of the protagonist and his (varying) girlfriends are made visible. The graphics are far from serious pornography: rather clumsy, mostly clothed, polygon people bump into each other. The reactions to the Hot Coffee patch distributed on the internet were conclusive. In the US the game was immediately rated AO (adults only) and thus in effect removed from the market, and in Australia it was explicitly banned. Subsequently, Rockstar

Games released a counter patch that prohibited activating the patch and newly delivered games refused the activation.

The implementation of controversial violent games widely follows gender-specific logics. The use of violence on unarmed women has, to our knowledge, not been a compulsory part of any missions to date. This also applies to the *GTA* series, although the game optionally allows players to attack or kill women.

Meanwhile, female characters who use violence themselves are mainstream. Violent women can also be found in *GTA San Andreas*. In the rural 'Badlands' missions, CJ and his Latina girl friend Catalina go on raids à la Bonnie and Clyde. And CJ's first girlfriend, Denise, is involved in drive-by shootings with hostile gang members.

The obvious imbalance between the absent discourse on violence and the public controversy about sexuality in *GTA San Andreas* is the more problematic since the depiction of sex/gender must be considered critically. To a great extent women in *GTA San Andreas* are treated as goods. Visiting a prostitute and purchasing her services raises the protagonist's 'state of health'. Furthermore, additional missions allow the player to fill the role of a panderer. The portrayal of the game's female characters is distinctly sexualized and, to follow Laura Mulvey (1975), presents them as objects of the male gaze: most women appear to be hardly older than 30, in most cases wear skintight, figure-accenting clothing and are slim. This one-dimensionality of the visual appearance of women in the game matches the conception of those six girlfriends with whom CJ has relationships. Entering a relationship in the first instance requires the so-called 'sex appeal' which is calculated from, among others things, cool clothing, a hip car, and a body shaped by body-building. At the same time female desire is constructed in a way that is quasi-automatically derived from questionable staging strategies of black, sexualized masculinity in terms of a simple stimulus-reaction scheme. No image of active female sexuality is depicted but permanent female sexual availability is suggested. To take control of communication is not possible. *GTA San Andreas* is far from being a differentiated relationship simulator. However, the various girlfriends do not react according to the same standard formula. The player has instead to adapt to different preferences in order to be able to enter a (sexual) relationship. On the one hand, there are 'classical' romantic concepts such as amusement rides or going out on dates. On the other hand, this romantic 'plot' is mitigated by Denise's demand for drive-by shootings or Millie's preference for sadomasochistic practices.

In the articulations and discourses of popular culture, romance is a concept by which the relationship of the sexes and the depiction of social minorities are represented as 'normal' and conforming to social common sense. Romance has to be classed among the complex cluster of heterosexual practices which, according to Lauren Berlant and Michael Warner (1999, p. 359) get

> confused, in heterosexual culture, with the love plot of intimacy and familialism that signifies belonging to society in a deep and normal way. Community is imagined through scenes of intimacy, coupling and kinship. [...] A whole field of social relations becomes intelligible as heterosexuality, and this privatized sexual culture bestows on its sexual practices a tacit sense of rightness and normalcy. This sense of rightness – embedded in things and not just in sex – is what we call heteronormativity. Heteronormativity is more than ideology [...], it is produced in almost every aspect of the forms and arrangements of social life: nationality, the state and the law, commerce, medicine and education, as well as in the conventions and affects of narrativity, romance and other protected spaces of culture.

The representational logic of the politically incorrect does, however, not consider the deconstruction of heteronormativity. The attraction value of the politically incorrect rather lies in the efforts to frustrate romance by means of degrading comedy and/or a combination of violence and romance, in particular at the expense of the social minorities involved. The latter is due to the regressive anti-PC logic of the controversial and the provocative.

As mentioned before, the construction of the semantic connection of race and sex in the game's central protagonist (and that of other African-American characters) more or less conforms to those exchangeable (self-)staging strategies of a male-centered African-American 'ghetto' culture as they are constantly reproduced in many hip-hop/rap videos. In these videos, women are almost exclusively featured as *rap candy*, scantily dressed and devoted to the singer. Similar strategies apply to the hyper-masculine stagings of, for instance, heavy metal videos. The sexualized and sometimes pornographic aesthetics of hip-hop videos, however, are quite different from the rather amusing Hot Coffee patch, even if the latter shows and enacts explicit sex. Furthermore, the meaning of sex and relationships, respectively, is, as in most computer games, secondary to the game's other missions. The underlying scheme – the construction of African-Americans as sexually omnipotent and the debasement of women as disposable sexual objects – are similar in both cases. The same applies to the heterosexual order, which, in the game as well as in many other pop-cultural representations of black ghetto culture, remains unchallenged and persistently bypasses alternative pluralistic concepts of gender and sexual orientation. Accordingly, CJ's sexual orientation is, despite all its modifiability, an unchangeable attribute. A homosexual relationship is not possible.

Conclusion

In many current Hollywood movies as well as in other cultural media forms, key concepts such as sexuality, romance and family are adopted to depict the 'normality' of social minorities. Those representations can be read as the attempt to establish political correctness, whereby political correctness here means the perceptibility of textual strategies of a (seemingly) non-discriminatory depiction of social minorities. Self-declared PC-opponents exploit such representations as alleged proof of PC's powerful impact in terms of a censorship authority that dominates culture and the *Zeitgeist*. The existence of such culture phenomena subsumed under the PC label, as well as hegemonic-conservative ideological concepts of this PC construction, form the background or quasi-inverse foil against which cultural strategies of the politically incorrect are articulated even if they are not explicit or adopted reflectively. According to the case study, *GTA San Andreas* is a convenient example.

If political correctness discourses can be conceived as an effective influence at all, then they can be only in respect to their imprinting, if not permitting, the feigned subversive nature of pointed taboo breaches at the expense of sociocultural minorities. Otherwise those taboo breaches would be nothing more than what they are: marketing techniques and pseudo-provocation that are no new phenomenon. In our analysis, we did not find indications of an ideological programme (or a message bound to specific social discourses) that was hidden behind the game's sex, gender or race constructions, waiting to be deciphered (and obviously distinguished from hate games such as *Ethnic Cleansing* that do not call for deciphering at all). For the most part, the provocative transgressions of boundaries serve to evoke the players' resistant and unattached pleasure and thus allow them to gain a feeling of distinction. This becomes clearer when one considers that such 'resistant' game modes (for instance cheats, the Hot Coffee patch) have already been implemented by the game's developers. Nevertheless, games such as *GTA San Andreas* cannot be regarded as harmless entertainment. *GTA San Andreas* and its specific connection of race, sex and gender consciously operates with negative stereotypes. Stereotyped constructions of social minorities as a structuring principle of cultural industry form self- and extrinsic images of cultural identities and lead to potentially dangerous exclusion processes as well as to an ideological justification of social inequality based on cultural difference.

Notes

1. This does not apply to the use of the term 'politically correct' within social movements such as Black Power or feminist activists of the 1980s. In this context, terms like 'politically correct/incorrect' did indeed bear programmatic features, though, rather in terms of an approach to the respective group's/ movement's self-determination (see Manske, 2002, pp. 29–36).
2. The extreme right-wing US website 'Aryan Knights' on Stormfront.org unsurprisingly enthusiastically praised *GTA San Andreas* for displaying a 'realistic' image of African-Americans, since the game showed how 'negroes have corrupted our society' (Leonard, 2006, p. 87).
3. See website: http://www.kyynel.biz/commando/.
4. For a list of these games, see: http://www.resist.com/racistgames/.

16
Strip – Shift – Impose – Recycle – Overload – Spill – Breakout – Abuse. Artists' (Mis-)Appropriations of Shooter Games

Maia Engeli

What started as a hacker attitude of writing patches to modify shooter games has developed into a genre of artworks, artistic endeavours that use shooter games in a variety of critical ways or that contextualize them in novel variations. These artworks and their special attitude of (mis-)appropriation are enriching the complex discourse around the shoot-em-ups with staggering aspects, ascertaining the inconclusiveness of the whole discussion by demonstrating a cunning flux of issues for consideration. In the following essay this field of artworks will be presented from a design perspective, framing it within eight different modes of de- and re-construction: *strip, shift, impose, recycle, overload, spill, breakout,* and *abuse.*

Idealistic worlds

Computer games, compared to traditional, physical games, have allowed a return to the archaic, archetypical, morally disputable pleasures of shooting, fighting and conquering. What started with *Pong* (1972), *Breakout* (1976) or *Pac-Man* (1979) developed into ever more realistic-looking realms and fights. The God view of the early games could be replaced with the immersive first-person view, thanks to the development of fast 3D graphics; a perspective predestined for aiming and firing, which is now happening in impressively rendered and populated worlds.

The rhetorics applied in the discussion of shooter games go beyond the seven basic play rhetorics of 'progress, fate, power, identity, the imaginary, the self, and frivolity' that were identified by Brian Sutton-Smith in *The Ambiguity of Play* (1997), where he demonstrates the contradictory nature of theories of play in the light of this variety of rhetorics. He points at a broader range of rhetorics that 'derive from beliefs about religion, politics, social welfare, crime and morality – that is, from all the matters that priests, politicians and salespersons constantly harangue folks about' (Sutton-Smith, 1997, p. 12). These rhetorics are the ones now predominantly applied in the discussion of first-person shooter games and they carry a great potential for confusion regarding the blurring of reality and virtuality in game worlds and game actions.

From the players' point of view the worlds are the primary reality where the game happens, where adventures are created, where fights take place, and where other events enrich the experience. The game world has to correspond to the nature of the game and afford the basic user interactions, allow the development of strategies, be supportive of the goal of the game, provide an adequate level of challenge and offer an interesting visual landscape. In earlier games, the worlds were built in ways to show dramatic settings while being very tolerant towards the (low) computing power necessary to render them. Early examples therefore displayed medieval-looking dark worlds with thick solid walls, dramatic lighting, cleverly placed hiding spaces, and heavy gates that open when triggered. As computing power and algorithmic cleverness increased, new approaches could be taken, such as futuristic-looking outdoor worlds with endless landscapes, complex architectonic structures, organic forms and a growing palette of sophisticated behaviours. But, regardless of the amount of computing power, the design of the world first of all has to be ideal for the game-play.

Experienceable subversions

> 'Art, like games, is a translator of experiences.'
>
> (McLuhan, 1997, p. 242)

The popularity of shooter games and the possibility to modify them has inspired gamers, artists, advertisers and scientists to create their own game worlds according to their respective interests. Among them are a number of new-media artists, who, since the mid-1990s, have created works by deconstructing, decoding or recoding shooter games. Anne-Marie Schleiner (1999) labels them as 'culture hackers who manipulate existing techno-semiotic structures' and Tillman Baumgärtel (2004) positions them in the larger context of 'the modernist art movement, to which they contribute some of the most sustainable ideas through the mis-appropriation of aesthetic ready-mades'.

The 1999 online art show 'Cracking the Maze – Game Plug-ins and Patches as Hacker Art', curated by Anne-Marie Schleiner, offers a valuable repository of some of the first works of this media-art genre, including a number of texts that situate its importance. The patches in the exhibition were built for games such as *Marathon Infinity* (1996), *Unreal* (1998), *Quake* (1996) or *Tomb Raider* (1996). Most of the 14 pieces in 'Cracking the Maze' addressed critical issues of the shooter-game genre per se, but not by disrespecting the fun-factor of games, as Erkki Huhtamo (1999) observed: 'The political determination should not be overemphasized. Humour and parody are important motives; the game patch artists don't seem to believe in the politically correct position of suppressing pleasure.' For example, in *SOD* the removal of information content transforms the game into a barebones, grey-noise state of itself and becomes an interesting parody of McLuhan (1997, p. 243) who declares 'the form of any game is of first importance' and sees 'the information content' as 'the noise and deception factor'. Others, including the epileptic virus patch or the female skin patch, more obviously brought up critical issues, such as the psychological tension created, and the brutal, overly male player figures and the macho codes dictating their behaviour. *Biotek-Kitchen* combined the game with issues of biotechnology to create a disturbing impact through the artwork. And in *ada_lovelace vs. donkeykong* 'the player interacts with the code itself' (Schleiner, 1999). Here, the shooter-modding art merges with software art and questions the significance of code and artificial machine language.

The breadth of issues in artistic game modding has increased since its beginnings, and so has the sophistication of the mods. There is an observable tendency away from focusing on the introverted discussion of issues related to the game per se towards an extroverted attitude of utilizing the game to create messages about issues external to the game. The idealistic quality of game worlds is also a precondition for game art. The artworks are based on an existing game or game genre, so therefore the actual artistic creation is experienced relatively, as a perversion of a known idealistic form. Nonetheless, through their work, the artists expand the breadth and depth of the discourse of shooter games.

Modes of deconstruction – a taxonomy from a design perspective

strip, shift, impose, recycle, overload, spill, breakout, and *abuse* evolved as categories in my 'Taxonomy – From a Design Perspective – of Art Game Mods of a Shooter Game or Art Shooter Game Developments' (Engeli, 2005). This taxonomy is the result of an analysis of this art form, based on an initial collection of 45 pieces selected because they were part of an important exhibition, mentioned in an important theoretical essay, or featured on a website that convincingly documented the importance of the work. In addition there are seven examples included that resulted from workshops I had been teaching in art, architecture and media schools.

An analysis and discussion of artworks usually focuses on the message created and the means used. My approach takes a design perspective aiming at understanding the artistic approaches taken to alter the games. The motivation for this approach originated at a series of game-modding workshops, where I needed a means to introduce students to the possibilities of creatively changing the game without predetermining the possible messages they can create.

The analytic process started with selecting works that would fit my primary criteria of being an important contribution to the shooter-game artform. The collection showed clear differences between extreme works, like the reductionist *SOD* or the overloaded *Nybble-Engine*, but generally the works could not be positioned as easily and the question was whether the analysis should take the form of a matrix, a cloud or an interactive multidimensional information-space. Since the goal was to create an instrument for the discourse about these works, taxonomy seemed the most adequate form, because it includes the creation of a vocabulary – a linguistic means that could be used to refer to a specific design attitude. The creation of categories through grouping the works and the naming of categories happened coincidentally. Only few works fit precisely into one category. Therefore the final mapping focuses on the predominant design attitude of each example.

The names of the categories are deliberately chosen to reflect the act of (mis-)appropriation, an act of (mis-)using the original in a more or less acceptable manner. It is an act of designerly deconstruction that breaks something apart and then constructs a diverging entirety. The taxonomy of *strip, shift, impose, recycle, overload, spill, breakout,* and *abuse* emphasises the destructive or violating part of this process.

strip – Reduction and intensification

Untitled Games, 1996–2001, jodi
SOD, 1999, jodi
Epileptic Virus Patch, 1999, Parangari Cutiri
ASCII Unreal, 1999, Vuk Cosic
Ah_Q – A Mirror of Death, 2004, Feng Membo

strip refers to the act of taking something away from the game on the visual and/or geometric level. A work in this category turns the remaining fragment into an impressive, intensive experience reduced to selected essential characteristics of the game. Most of the examples in this category are from the early days of shooter-game mods, the time when the artistic discourse was focused on the game per se. A prominent example is jodi's *SOD*. Antoinette LaFarge (2000) describes it as follows:

> Their game *SOD* takes the standard structure of an action game [...] but changes the usual literal architecture to an abstract world of black, white and grey planes. *SOD* is highly disorienting as a result; not only is there no visible difference between floors, walls and ceiling (which way is up?), but it is very strange to find oneself menaced by geometric abstractions (what *was* that I just shot? is it even shooting anymore?).

A newer example is *Ah_Q – A Mirror of Death* by Feng Membo. He emphasises the self-reflective quality of shooter games through the replacement of the walls of a simple rectangular game space with mirrors. All the players look the same and are modelled after Feng's own appearance and equipped with video cameras to record everything.

shift – Visual reformulations and the power of the look

Nude Raider, 1998, Rob Nideffer
Hotel Synthifornia, 1999, Fuchs-Eckermann
Pencil Whipped, 2000, Lonnie Flickinger
SkinPack, 2001, Linda Erceg
cuteXdoom, 2004, yumi-co

The *shift* category contains artworks in which the visual appearance of the surfaces of the shooter game world or the avatars has been altered. By just changing the look, without modifying the game's underlying mechanisms, these examples achieve a considerable change of the perception of the game's substance. Manipulating the surfaces is thus a very efficient strategy for manipulating the meaning of a game. The first *shifts*, like the famous *Nude Raider Patches*, were not done by artists but originated from game-hackers. They became the basis for an artwork when the artist Robert Nideffer patched those patches by adding a moustache to the nude Lara Croft. '[They] offer alluring variations on Lara; transsexual Lara, butch Lara, Lara in drag ...' (Schleiner, 1999) and added a new dimension to the discourse of computer games and gender issues. In *Pencil Whipped* all surfaces are replaced with sketch-like textures. As a result the whole game, including the players, looks like a sketch of a game and it seems impossible to play it in the same sincere manner as the fully detailed game. Also *cuteXdoom*, the chill-down spaces of *Hotel Synthifornia*, and the armed, fragile, naked women of *Skinpack* demonstrate the power of the look on attitude.

impose – Insertion of a cultural, artistic, political or social occurrence

Biotek Kitchen, 1999, Josephine Starr, Leon Cmielewski
Alice, 2000, America McGee
Through the Looking Glass, 2000, Miriam Zehnder, Eric van der Mark, Patrick Sibenaler
Quake Friends, 2003, Joseph DeLappe
2004 Presidential Debates (The Great Debates), 2004, Joseph DeLappe
Suicide Solution, 2004, Brody Condon

The examples in this category use the game's structure as a framework within which they reference, for example, a story, a TV show or a political event. The category is named *impose* because the game becomes the platform for something that is foreign to it. The imposed events have some fame to them; there is *Alice in Wonderland*, *Friends* (the sitcom) or the *2004 Presidential Debates*. The citations have to be at least roughly known to the player so that the superimposition with the game can create the tension that leads to a change in the perception of the imposed event. For example, *The Great Debates* are re-enactments of the three 2004 US Presidential debates made by typing the transcript of them into the discussion interface of a game. It took the artist, Joseph DeLappe, about eight hours for each performance. His documentation reflects some of the interesting moments of the dialogues with players, who did not know what they were involved in. They seemed to get annoyed and then give each other hints like 'type ignore 14 in console to ignore Bush'. There were also moments when the players contributed a few of their own comments to the debate. While the *shift* attitude changed the perception of the game, the *impose* attitude resorts to leaving the game more or less intact to create strong messages in relation to the imposed theme.

recycle – Citations of real world forms and locations

Ars Doom, 1995, Orhan Kipcak, Reinhard Urban
Museum Meltdown, 1996–99, Thomas Bernstrup, Palle Torrson
Therme Vals, 1999, Miriam Zehnder et al.
Dreamday, 2002, Marc Dietrich, Michael Huber
911 Survivor, 2003, Jeff Cole, Mike Caloud, and John Brennon

America's Army, 2002, Michael Zyda, et al., the MOVES Institute, the Naval Postgraduate School
Chinatown, 2002, Brody Condon, with Eric Cho and Sky Frostenstein
Vietnam Romance, 2003, video Eddo Stern
Waco Resurrection, 2003, Michael Wilson, Eddo Stern, Jessica Hutchins, Brody Condon, Peter Brinson, Mark All
Escape from Woomera, 2004, Kate Wild, Stephen Honegger, Ian Malcom, Andrea Blundell, Julian Oliver, Justin Halliday, Matt Harrigan, Darren Taylor and Chris Markwart
Dollhouse, 2005, Maia Engeli et al.

In this category, the game's virtual world is replaced with replicas of something that exists in the physical world. While in sports games such a coincidence is desirable, shooter games tend to provide rather distinct virtual worlds. Playing a shooter game in a virtual 'real world' replica has an effect on the way the physical place is perceived thereafter. For *Therme Vals* the artists remodelled a spa – an architectonic masterpiece by the Swiss architect Peter Zumthor – by precisely capturing its surface structures, light situations and sound qualities and offering it as an environment for a shooter game. After playing this game the spaces of the spa are not the same, and the virtual events of the game become part of the history of events related to it and thus become part of the experience of this place. *Escape from Woomera* is a very powerful, politically oriented work. It is a remodelling of the Australian refugee detention camp at Woomera and makes it possible to

> live through the experience of a modern-day refugee in the most secret-ive and controversial places on the Australian political and geographi-cal landscape. [...] It is no surprise given these conditions that refugees in detention, like many others unjustly imprisoned before them throughout history, routinely stage heroic and sometimes very success-ful breakouts from captivity in their fight for freedom. (Wild et al., n.d.)

It is this experience the artists want to make accessible, not how to escape. The *recycle* category has some similarities with the *impose* category, but here future encounters with the physical space will trigger the memory of the actual game experience, thus *recycling* the experience.

overload – **Superimpose the game mechanics with other processes**
LinX3D, 1999, Margarete Jahrmann, Max Moswitzer
Adam Killer, 1999–2001, Brody Condon

Velvet Strike – Counter Military Graffiti, 2002, Anne-Marie Schleiner
qqq, 2002, Tom Betts
Nybble-Engine and *Nybble-Engine-ToolZ*, 2000–2004, Margarete Jahrmann, Max Moswitzer
Dancemachine, 2003, Margarete Jahrmann, Max Moswitzer
fluID, 2003, Fuchs-Eckermann
Desimulat, 2004, Valentina Vuksic
Stunt Dummies, 2004, Kathleen Ruiz (GameArt)

overload refers to the expansion and addition of processes. The original shooter game is still playable, but other behaviour has been combined with it and interferes with the familiar operation of the game. Due to the different overloaded processes, the game's actions can become very complex and multifaceted. In *Velvet Strike – Counter Military Graffiti*, CS 'sprays' (graffitis) can be up- and downloaded through the internet to be sprayed onto the surfaces of the online multiplayer terrorist game *Counter-Strike*. Furthermore, specific tactics for players to undermine the game, so-called intervention recipes, are provided. While the processes of the game engine are not altered much, the game is manipulated by motivating a player community to invade the game with new tactics and 'weapons'. In examples such as *qqq* or *LinX3D* and *Nybble-Engine-ToolZ* the actual internet communication processes of the game have been altered to access more of the data generated through networked play, to use this data in generative processes, and to send out messages over the internet (for example, anti-war messages to president@whitehouse.com in *Nybble-Engine-ToolZ*). The adding and overloading of processes demonstrates the potential of the game engines beyond their graphics qualities. There are numerous other aspects, mainly behaviours and networking processes, which are accessible for modification and offer a wide field for further artistic exploration.

spill – Expand into the physical realm

Consoles, 2001–2004, Paul Johnson
Cockfight Arena, 2001, Eddo Stern, Mark Allen, Jessica Hutchins, Karen Lofgren
Tekken Torture Tournament, 2001, Eddo Stern, Mark Allen
Can You See Me Now?, 2001, Blast Theory
Fort Paladin, 2003, Eddo Stern

The *spill* instances add a physical context to the digital game, like specifically designed game consoles, the provocation of real pain in players or the extension of the game world into physical space. The game *Can You*

See Me Now? happens simultaneously online and on the streets. Outdoor runners, supported by GPS, walkie-talkies and handheld computers, have to catch the online players, who can see their pursuers on the screen. The creators emphasize presence and absence as well as distance and proximity as major themes in this work and add 'the virtual city (which correlates closely to the real city) has an elastic relationship to the real city' (Blast Theory 2001). But games are also about harming and killing the enemy, an aspect emphasized in *Tekken Torture Tournament* where the fighting players feel actual pain caused through non-lethal electric shocks. Both *Fort Paladin* and *Consoles* are about computers fighting against each other. There is no role for a human player and the game becomes an autonomously running installation. Without human players it seems like an amusing parody of a shooter game and the question is whether it is about violence as pure entertainment or a reminder that machines can fight better than humans. In these two examples the human cannot be a player immersed in the game world, but only a spectator from the physical realm.

breakout – Outside the digital realm

Shooter, 2000–2001, Beate Geissler, Oliver Sann
Unreal Overall, 2004, Shusha Niederberger
650 Polygon John Carmack v2.0, 2004, Brody Condon
Untitled War, 2004, Brody Condon
MÄDunreal, 2004, Synes Elischka
You're Dead, Game Over!, 2004, Yan Duyvendak

The *breakout* examples escape the digital realm and manifest themselves in the physical space either as recordings or interpretations of the whole game, or as specific aspects such as the shooting action or the geometry. *Shooter* is a series of portraits of players engaged in a game. The tension created through their involvement with the game can be sensed from their facial expressions. *650 Polygon John Carmack* is a CNC milled sculpture of a low-polygon version of the founder of ID Software, the Texan computer game company. But the sculpture not only honours John Carmack – 'who together with John Romero was among the first to openly provide the code of their games for modding' (Baumgärtel, 2004) – it also demonstrates impressively the abstractions that games work with. An unusual work in this category is *MÄDUnreal*, a board-game version of the shooter game *Unreal Tournament* with frags, weapons, health packs, a forking course, teleporters and everything else necessary for good game-play. In the *breakout* examples, reflection always happens on two levels, as a direct reference to the game, and as a critical reference by exposing the virtual in the physical. The *breakout* attitude also adds a nice tactility to the works, which contrasts with the remote interaction in computer games; therefore *breakout* can also be understood as an escape into a more haptic, tangible realm.

abuse – Neglect the shooting

q3apd, 1999–2000, delire (Julian Oliver)
Quilted Thought Organ, 2000, delire (Julian Oliver)
Expositur, 2002, Fuchs-Eckermann
Studio 7, 2003, Marc Champion, Matthias Branger
deRez_FX_kill<Elvis, 2004, Brody Condon

Shooter games offer a wealth of possibilities for the creation of dynamic and intelligent virtual environments with high-quality graphics and sound. The *abuse* examples exploit a shooter engine to create works such as virtual exhibitions, instruments, interactive videos or visuals for audio performances that leave out the shooting. These examples do not actually contribute to the discourse of shooter games, but rather to discourses of the art forms they serve; for example, digital music and visuals. *q3apd* and *Quilted Thought Organ* are virtual instruments that also produce visuals. This combination puts new demands on the performer, because the same action results in audio and visual output simultaneously. *DeResFX_Kill<Elvis* shows a way in which the shooter game may be reflected in works of the *abuse* category: 'As the viewer camera floats through an infinite pink afterlife, twitching multiples of Elvis are controlled by the original game's "Karma Physics" real-time physics system – generally used to simulate realistic game character death' (Condon 2004).

In this overview of the taxonomy, only a small number of examples could be described. Currently there are 52 works in the online version of the taxonomy. The selection will be updated whenever additional relevant new works are found. The taxonomy can be visited at: http://maia. enge.li/gamezone/taxonomy.html.

Beauty and the beast

The beauty of this art genre is that it is about experiencing the game. This art is not created from scratch and consequently is hard to understand without some previous knowledge of the game. It is offered through a de- and reconstructed, de- and recoded world that one has to engage with. The art of (mis-)appropriation has to be understood as the creation of experiences. The design attitudes reflected in the terminology of the taxonomy – *strip, shift, impose, recycle, overload, spill, breakout,* and *abuse* – refer to the destructive aspect of the process.

To focus on the design aspect, within the discourse of shooter games, may seem like a retreat into less risky territories that avoids burning questions like gender issues or violence. Artists face such issues. Therefore, in addition to the design attitudes, there are also indications of how the game is exposed in a new way.

The taxonomy aims to serve as an inspirational instrument for the creation and discussion of artistic shooter-game mods. It has been referred to by others such as Domenico Quaranta (2006), who perceives *abuse* as a very 'efficient expression' for the categorisation of *q3apd*, or Cynthia Haynes (2006), who was inspired to create a taxonomy of 'god modalities' for *Disarmageddon Army* (a mod of *America's Army*) to structure concepts and create a reflective terminology.

Part VI
Computer Game Play(ers) and Cultural Identities

17
Presence-Play: The Hauntology of the Computer Game

Dean Lockwood and Tony Richards

In contrast to many theorists in the growing field of computer game theory who point out the strong 'apparatus' nature of the player-game interface (an ultra-hypodermic experience; a cybernetic loop), we foreground the active and non-immersive nature of the game-space. Film theorists have often tended (ultimately under the influence of Althusser, 2001) to posit a passive identification with the filmspace (for example Baudry, 1985). Althusser argued that media, as 'ideological state apparatuses', reproduce the social order through securing our unconscious consent. In this view, cinema has a political significance as a system of representation offering identifications enlisting spectators into pre-organized subject positions (a process Althusser termed 'interpellation'). This already suspect one-to-one relation has been uncritically imported into computer game theory whereas, conversely, identity-problematization, undecidability and play are integral to this new medium.

To serve as an exemplar of the Althusserian influence in computer game theory, let us dwell briefly on Matt Garite's paper presented at the Level Up conference in November 2003. His paper, 'The Ideology of Interactivity', presents in stark fashion some of the presuppositions typical of the Althusserian critique of the computer game. His thesis is that the true *raison d'être* of games is to work us unto death. The 'political unconscious' (Jameson) of the form is directed towards the Taylorization of leisure in the digital age. 'The command structure of videogames,' Garite (2003, p. 12) says, 'tends to reinforce the disciplinary regimes of late capitalism'.

In this view, and against the hype of interactive choice, the work games do is to 'play' the player: 'By repeatedly demanding user input, video games lock players in a self-replicating, integrated circuit of instructions and commands' (Garite, 2003, p. 2). Garite, following Manovich (2001), suggests that computer games embed a Brechtian aesthetic which serves to heighten the user's awareness of the machinery. However, this insight is entirely short-circuited by his conclusion that the 'auto-deconstructive' alienation effect of games arising from the player's repeated forced shift between the roles of active user and passive spectator (as in the transition from game-play segment to cutscene) does *not* make them available for critical engagement. In fact, the oscillation actually serves to suture the player more securely into the game (Manovich, 2001, p. 208). As Manovich argues, we are presented with a *fait accompli*, a thoroughly enclosed structure which does not permit any 'filling-in' of consequence by the player. Such a structure does not even permit the kind of psychological interactivity afforded by older media such as cinema (p. 61).

In short, computer games 'operate on players through an updated, aggressively interactive and immersive form of interpellation' (Garite, 2003, p. 5). The surface interactivity interpellates the player as a 'freely acting individual' but any options are already determined by the deep structure, the hidden code (p. 6). This is serious stuff – the game performs a symbolic violence on its player. Like a virus it replicates itself in player behaviours, merging the player with code (p. 10). Computer games are 'weaponized texts' and 'living rooms and bedrooms are now occupied territories' (p. 8). In this cynical vision of game interactivity as the digital age's 'Arbeit macht frei', gaming becomes a coercive prod to the terminally lazy, compelling them to take up their positions at the PlayStations (or workstations) of the world: 'gaming is essentially an aestheticized mode of information processing' (p. 10).

Garite argues that, 'like cinema, the videogame seemingly permits us to wander while it chains us to our seats' (p. 7). This chimes with Bolter's (2002) argument that games remediate cinema. Games borrow or refashion the formal characteristics of cinema and, along with this, the ideological subject-positioning work of the cinematic apparatus. This exacerbates the spectator's identification with the hero, providing an even 'more effective means of co-opting the user into the ideology represented by the game' (Bolter, 2002, p. 86). Computer games, with their powerful combination of immersivity and interactivity, *extend* the power of cinema to interpellate the user. They remediate the cinematic gaze, particularly, it is argued, in the deployment of first-person point of view. Thus, Mulvey's well-known points (1975) about the surrogacy of the protagonist for the spectator are painted here, in the game-space, on a much grander scale. By giving power to the gamer in the form of controls, by immersing the sensory-motor engagement of the player 'into' the game-space, the player becomes a first-person subject of the space. This paradoxically robs the player of conscious awareness or critical distance. They are fashioning their own text as they render themselves powerless.

To return to Garite, then, 'the world of the videogame is nothing more than the on-screen rendering of programmed instructions and decrees' (2003, p. 8). We strongly disagree with the way such statements recuperate games for a respecified vision of old-style interpellation, jazzed up in the terminology of new media.

Žižek (2000) has noted how, comprehensible only after the fact, established media struggle to capture unfamiliar experiences, accommodating them in 'excesses' which stretch formal boundaries. Thus, much of the radical experimentation in the classic realist novel now makes sense as an attempt to articulate phenomena more conducive to cinematic representation. Žižek gives the examples of Emily Brontë's 'flashbacks' and Dickens' foreshadowing of parallel editing. Today, he argues, similar excesses occur in cinematic narrative: 'a new "life experience" is in the air, a perception of life that explodes the form of the linear centred narrative and renders life as a multiform flow' (p. 39).

Along the same lines, Kinder (2002) has identified a phenomenon which she dubs the 'game film'. She discusses a number of experimental films in the 1960s and 1970s by directors such as Buñuel, Resnais, Marker and Varda, all of which embed games at their centre, and suggests their appearance reflects a dialogue between filmmakers and newly emergent poststructuralist approaches (p. 128). The idea of the game film is useful in helping us gather together the more recent explosion of playful experimentation within cinema, of which perhaps the best example is the work of David Lynch. His work is often bundled in with the kind of ironic pastiche of mediatized society associated with postmodernism. However, where the sly digs perpetrated by Oliver Stone's *Natural Born Killers* (1994) and other self-conscious postmodern narratives cannot do without the security of having the subject in on the joke, Lynch's movies set out to cross-fertilize Hollywood linearity with avant-garde techniques of estrangement in ways which do not permit the audience any immunity. Intermittently incoherent in narrative, character and space, they deliberately set out to throw audiences into crisis without offering any recuperative position (Rombes, 2004, pp. 74–5). They are part of the deconstructive sensibility we wish to highlight, committed to an exploration of the indeterminate nature of identity. In Lynch's *Lost Highway* (1997), the protagonist inexplicably metamorphoses midway through the movie into someone else. Although it is sorely tempting to solve this puzzle by imposition of a some traditional film-studies framework such as psychoanalysis, this hardly seems in the spirit of Lynch's game, which is not to teasingly hold out on the information we need to answer the question, 'Why?', but to frustrate identification and the formation of a secure viewing position for the audience at every turn. The movies bear witness to 'the chaos of contradictions [...] that go into the making of any self' (Rombes, 2004, p. 75).

The game film testifies, as Žižek (2000, p. 40) supposes, to a perception of reality as haunted by other possible outcomes. Reality is rendered more fragile, ineradicably contingent, as traces of other paths, simultaneous possibilities for identity, are introduced. Cinema is pushed to the limit trying to articulate this perception and, just as it picked up those earlier threads thrown out by the novel's excesses, so the computer game (not, *pace* Žižek, the hypertext) picks up these new threads and makes them its basis.

If play was a marginal force in old media, it is without question a central part of new-media cultures (Dovey and Kennedy, 2006). The conceptualization of play for new-media theory, some have suggested, should foreground the riskiness of play and identify what precisely is at stake when we play in and with new media (Kattenbelt and Raessens, 2003). This gesture towards riskiness and reflexivity moves in the right direction but does not go far enough in overturning what we consider to be the identitarian framework taken for granted by almost all computer-game theory. It still suggests a subject secure enough to take risks. What is at stake, for us, is precisely the self-identical standing *of* the game-player.

Here then, *in* the game-space, we are not occupying a linear-textual mechanism that attempts to lock in and baptize spectator identity through a linear discursive-hierarchical structure (MacCabe, 1981), a rigid renaissance architecture that attempts to structure our *approach* and our final subject position through the hiding hand of narratological perspective. Far from games being a remediation of film (à la Bolter) we have instead a different kind of structure, not so violently hierarchical but also, as we will see, not as polymorphous or anarchic as certain other new-media cousins. Perhaps then there is a third way here; vibrating, oscillating and flickering between the other two: a presence-play.

It would be important to expand on what *we* mean here by play. It is not of course in the commonsense 'game-playing' sense that we are talking. This play is very serious. Nor is it 'playing' in the Bakhtinian 'carnivalesque' sense of the word (a ritualistic resistance to authority and structure), nor even the de Certeauan sense of taking what is given and making of it something else. It is more *uncertain* than these. It is play as an *always* unavoidable putting into question of notions of 'Presence' in the sense that Derrida has continually tried to redefine this problematic.

In applying this to the computer game we are pointing out that, unlike within linear apparatus theory, we are here arguing for the *objectifying* of the medium of computer game as an identity-*challenging* space, an identity-'*un*deciding' space; a '*playing*-machine-medium'.

It is true of course that deconstruction has been taken up once already by a member of our 'new-media' fraternity. These deconstructive concepts have been taken up and made to function for what could be called the post-human, 'post-text' or hypertext. Such theorists of this space, as we will see, tend to believe that in their object (their 'representative' for which they are the theory) they have now side-stepped the 'challenge-to-identity' and have now stepped into a new chapter: the web. Are presence and identity, however, so easily cast aside or exited? To quote Derrida (1967, p. 280):

> The step 'outside philosophy' is much more difficult to conceive than is generally imagined by those who think they made it long ago with cavalier ease, and who are in general swallowed up in metaphysics by the whole body of the discourse that they claim to have disengaged from.

Such theorists of the 'hypertext' believe they have now escaped such tyrannies of singular meaning (of presence *outside of* difference) by there now having recently been made practically available wormholes between *all* texts. But again what is it 'to escape' from identity? Is it so easy to disassemble presence? What exactly does the hypertext then overcome?

Western thought, as Derrida often points out, is constituted by a history of attempts to stabilize, to leave in place certain 'transcendental positions' that simultaneously exist outside of a system yet also orient and partake of it (being at the same time inside and outside, narrational and meta-narrational). Western culture has an arsenal of such orientation nodes that serve to pin and stabilize the fabric and to rule out play. According to Derrida (1967, p. 281) these centres: '[O]rganize the structure (one cannot in fact conceive of an unorganised structure) but above all to make sure that the organizing principle of the structure would limit what we might call the freeplay of the structure.'

These points of presence ('the origin and the end of the game'), these presences 'that attempt to escape play' (for example arche, telos, consciousness, patriarchy, national identity, God) are however false closures; they *hide* and try to tuck away certain niggling self-doubts; that the truth of their presence/being is actually at play. Deconstruction attempts to locate these areas of self-contradiction that destabilize and threaten the viability of a particular text (as exponent-child of the larger Western structure). This exposure (by deconstruction) is not taken as something 'from-the-outside' then but a locating of a point of self-critique, of *played* presence already there, but hiding (Derrida, 1989). This 'presence-belief', then, is not at all an easy *position* to escape and is then perhaps a necessary delusion (the hiding and tucking by Lévi-Strauss [1983], for instance, of the now exposed and famous 'incest prohibition' that puts into play the attempted 'presence' of the construction of the *seemingly* separable opposition between nature and culture).

A question might arise then: why attempt to construct some *'new'* texts or cultural spaces if those existing texts and cultural spaces always already do this within their margins anyway? Deconstruction is not just about critiquing such third-party texts but of creating certain playful new texts which attempt to play within themselves; to exhibit more *honestly* this play-of-presence. Not, it must be re-underlined, in a belief in having kicked away a ladder, but in having loosened a knot. The point in these spaces then is not to hide-away-from-play, to *'pretend-presence'* but to display and affirm this plasticity of being. It is not, however, to abandon the game to anarchy and escape to the belief of having escaped through the wormhole.

This is then no post-philosophical, destructive or anarchic game, as detractors and certain supposed advocates have accused or praised it of. Before going on to stake a claim for the game as identity *un*deciding machine it will now be worth looking in more detail at some of the claims that its new-media cousin, 'hypertextual theory', makes for being an advocate or baton-carrier for deconstruction.

One of the first theorists of hypertextuality was Landow. In a landmark book titled *The Hypertext* (1997), he put Derrida to work in helping to find a theoretical foundation for the quite new *post*-linear practice of hypertexting:

> [Derrida is] groping for a way to foreground his recognition of the way text operates in a print medium – he is, after all, the fierce advocate of writing as against orality – shows the position, possibly the dilemma, of the thinker working with print who sees its shortcomings but for all his brilliance cannot think his way outside this *mentalité* (Landow, 1997, p. 34).

Giants standing on the shoulders of dwarves. Leaving aside the excessive self-confidence in his deployment of the French language, this representation of Derrida (somewhat along the lines of an 'if *x* were around today s/he would be partaking of *y*' formulation) of course could do with some unpacking; unpacking assumptions on both what Derrida is here made to stand for and assumptions on how this hypertextual theory can develop and move Derrida forward.

In pointing out how certain concepts within deconstruction reach towards and fit very well within the practical workings of hypertexts, Landow points out that Derrida's search (*within* linear writing's confines) for a discourse to disrupt presence (to put those sureties like nationalism, God, etc. under question or erasure by their 'other-spectres', the trace of their Other) is in fact doomed to failure. Derrida is trying, but without really *knowing* it; without being able to understand the 'limits-of-his-vessel' and his own indirectly echoed dream for an-other vessel yet to come. Put simply he is using a writing technology not fit for the job he indistinctly wishes to be doing. However, in making this call that he does not know he is making, the hypertext has arrived in answer. Given this hypertext toolkit Derrida might have found the going a good deal easier, 'but hats off to him anyway'. This amounts to a fundamental misreading of deconstruction.

It is a fundamental misreading because he makes a move that Derrida does not sanction (even in those foreshadowings of his manifest dreams) and which misinterprets Derrida's 'latent' motivations.

Derrida speaks of a form of writing which *would* 'play' rather than, as we have said, simply overturn the tradition. In writing (or in *writing*) in a style that attended to language's slippery nature he was not however advocating *abandoning* writing to any concept of 'post-writing'; of a 'post-human' 'post-textual' or hypertext if you like. He was not suggesting abandoning texts and reaching out for 'the web' (of language/s), was not mixing up interpreting (as web, trace, gram) or mixing up writing (*playing* with those trace-like qualities) with some polymorphous expansive universe. Put simply (for heuristic measure) in constructing a 'parole' (as speech act) he was pointing *towards* the 'spectral' nature of these paroles in relation to a preceding 'langue' (that a concept is *caught-up* in a system-of-differences and cannot thus *escape* its difference; it is *'the trace'* of its difference). In constructing a 'parole' Derrida is not hoping to blend or unite into 'la langue' (a mystical unification with the other). Deconstruction is not *praying* to have 'constructed' some-such post-text, meta-text, web-of-texts, nor hypertext. To believe that one has moved *outside* of the text is to move into the non-space of the wormhole itself. This aspect of the hypertext offers a sort of anarchy of reading and of *écriture*.

There is a further aspect of the hypertext which contradicts somewhat this *supposed* Derridean dispersal but which Landow (1997, p. 33) simultaneously seems to celebrate also, and also somewhat giddily: 'Derrida properly acknowledges (in advance, one might say) that a new, freer, richer form of text, one truer to our potential experience, perhaps to our actual if unrecognised experience, depends on discrete reading units'.

Now, within the hypertext there seems to be present a barely hidden (pre) anti-deconstructive current that believes in an encircling or circum-navigating capacity within such spatial travels (Bennington, n.d.). There is the belief within hypertextual theory that paradoxically points towards a certain encyclopaedic quality, of having *more* knowledge, and therefore *more* presence and even more of one's self (Bennington and Derrida, 1993). This *encompassing*, this immanent networking can put us into contact with everything anywhere and, like our own brain, this new brain will allow us a clear and true and present speech. Very un-deconstructive.

Where then can the computer game fit between the violently hier-archical linear text and the violently anarchic post-text hypertext?

The game offers a space which is neither quite textual nor 'post-textual'. It neither closes off on a linear narrato-ideological position; a framing point of vanishing consciousness for the player. Nor does it open upon so many exits that it ceases to be a text. In the game we are not then text-ually conjoined to a textualized cipher/stand-in, an identificatory 'truth-spot' as it were, but to a digital text with paradoxically analogue-plastic qualities that, unlike in the stage-based 'digital' world of the analogue film, is lived on the basis of the hack. This hacking is not something that is done to the game-text from the outside, as a violation to the author-ial code or position, but is the permanent being-towards-the-game that the gamer comports him/herself within ontologically: a *being-at-play*.

This hacking presence is the non-escapable ontology of the game; it *provides* the game. It *makes* it playable. Without this play, if some future game *could* conceivably escape such hacking, the game would be a plain old return to film because nothing the player did could change the course the server-game had set in place ahead of time. The wish and the very real attempts to reach back to film, evidenced in all those ideologies of the linear and the accompanying fetishized photographic naturalisms (with the mere *seeing*-supplement of client-sided game-control) provide a telling and insurmountable contradiction. We must define and make clear our oppositional hack-play (hack reconceived as a non-problematic or *invite*); investigate the condition of possibility for this presence-play, and 'hacking' as simultaneous-play of presence-absence. What then of this reappropriated term hacking?

Hacking here should be re/appropriated or re/conceived as a pragma-tological *haunting*. As Derrida points out, haunting is not a hermetically closed chrono-historical step-by-step progression and thus does not take place upon some locked causal trajectory. There is then no mathematical future to hack into or de-rail within the game, no guarantee to prepare for; just a *spatio*-temporal haunting by other possibilities. Hacking plus haunting is an intervention within the 'text' which is *always* in contact with its other paths, wherever they are. Haunting adds a spatial modification to the unfolding temporal and thus also *un*decides that temporal. The place we are here *in* is haunted by a vacant position of *an-other* place in the game which is not at all *present*. It is a ghost of a place, an inescapable spectre hanging over the occupation of *this* place. Haunting-hacking provides no violation to any closed-up identity in the game when this place can be/is that place also. There can be no misdirecting of linear movement when the absent path is always *playfully* haunting in the 'present's'. No presence-in-the-present. Concomitantly to this, the present that we are *on* is not some *staging* toward a textual-transcendental final-identity, but the haunted ash or remainder of an-Other pathway, *over-there*. There is no room in the game to hide away from play. The game perhaps for this reason alone would not make a very powerful or suitable tool of nationalist propaganda and this points also to its deconstructive nature vis-à-vis identity stitching or suturing (Heath, 1981): a *différance* machine.

With these audience 'interactions' then not being a hack or a violation to the directionality or *focus* of the text, it is also important to reinforce that the freedom and anarchy of the hypertext is not being repeated on a narrower scale. The game-space is plastic but does not offer a slightly smaller infinite; play involves a hacking of the space but not a breaking of its boundaries. This is the game that is this ghostly-medium's presence-play.

Let us turn now to a brief concrete example, which we may not, however, be able to completely wrap our hands around. A game such as *Black & White* (2004) by Lionhead Studios, in contradistinction to its name, would take the form of an *analogued*-binary. The black and white of a filmic *binaried*-analogue is replaced in this game by a white that would *always-already* be contaminated by its blackened-Other. There is thus no *excluded* opposition/other but a continually *inhering*-haunt. The game, in providing user-control and range (of which *this* particular game is just a clearly ludic and lucid example), supplements something unavailable to the film (and which cannot be thrown away *later* by the game). It is a supplement not in the sense of what is *already there* but only *increased* but a supplement that transformatively violates, never to return (to any pre-ludological state). This supplement then, as we have seen, is not a mathematical addition *to* the film, a *more-so* or *Über-Film*, as it were, but a qualitative addition that turns any such thoughts inside-out; an *analogue-binary*. This inside-out of the closedness of the filmic binary-analogue (presence-identity) is once again not to repeat the 'all-exits' hypertext.

Here then, as in Žižek's championing of multiform experience, every choice that is made for the beast is explicitly haunted by a choice that was not, or is not, *'being*-made'. Here we return to the game's ontologically morphous *haunting-hack* (a hauntology) but not to the vexed issue of gamic-deaths or finality. Death is not to be mathematically guaranteed, as it's *upon* some other path in *some* future form. Derrida proposes two senses of a future which seem appropriate here when distinguishing between the singular 'humanist' future of the film and the haunted-decentred future that is the future of the game: a future already-scheduled and a future *un*-certain.

> The future is that which – tomorrow, later, next century – will be. There's a future [*le futur*] which is predictable, programmed, scheduled, foreseeable. But there is a future – *l'avenir*, to come – which refers to someone who comes whose arrival is totally unexpected. For me, that is the real future: that which is totally unpredictable, the Other who comes without my being able to anticipate their arrival. So if there is a real future beyond this other known future, it's *l'avenir*; in that it's the coming of the Other when I am completely unable to foresee their arrival (Derrida in Kofman and Dick, 2005).

The game's future path here is haunted by the spectre-come-alive-*of* its Other-path. A wafer-thin membrane wobblingly separates these spectral spheres. A very thin protection then; one path-position, spectrally, pollinated by the other. In truth one is founded on the other as of a present-absent spectre *impregnating* the present; practically and compossibly brushing alongside as in a ghostly shiver in our presence. This making unfolds itself as a *spatial*-difference to this Other-path. It is then the remaindered ash of an-other path, *over-there*. The present is never-presence outside of this haunting. The haunting then enfolds *as* it unfolds. Even once the central character of the beast, given game-time, takes-hold-of-an-identity and gains recognisable behaviour patterns formed as a relatively continual locus-of-operations; the trace of other choices, other beasts and other presents haunt and enfold as *différances*. If these *unfolding* hauntings were not enough in themselves, these hauntings are underlined and made transparent by a room in 'the temple' (a visitable truth tower, as it were) where 'stages' of the game can be *saved*. These recoverable moments or savings would seem perhaps to be *innocent* stage-saves, much like the obliterative savings of some word-processed document (where that previous saved-state is rubbed over leaving no ash). These saved stages are not merely temporal stages but spatial states also (Derridean *différance*): an irresolvable combination of the two (differ + defer). The temporal closure of the game will never come outside of this *différance*. There is no singular, carved or prepared-for death as there is no ongoing singular syntagmatic syntax. No *death-syntax*. Thus, to provide a rather obvious example and one that does no justice to the detailed complexity of the disseminatory effects of these *savings* (which are not singular savings but cover the wall of the 'savings-room' as many-haunted alternate stages), if we save a 'stage' where the beast has been treated brashly we will have a *différant* beast to the one saved in the temple that was treated upon an-other 'stage' in a gentler way. Within these *stages*, there can then be no singular and unavoidable death being lead-up to. No *absolute* stage.

Reflecting on the complexity of this, we must be wary of believing ourselves to be capable of embarking substantially on any 'finalizing' *act* of analysis. There is no 'concrete' object to place our tools *upon* (cleanly transporting subject-to-*object*, analysis-to-*analysand*, local-to-*global*). Indeed the status of the video game works, relative to the classic-filmic text, analogously to replicate the difference of classical 'object' physics (Newton to Einstein) to quantum physics (the particle/wave 'complementarity' of Niels Bohr [see Plotnitsky, 1994]). In this scenario we have the impossibility of speaking of there being any object *prior-to-playing* (or analysis) of *any* isolatable status: there is an *irreducible* impressional-touch that 'the encounter' makes *upon* 'the object'. The highly complex 'undecideability' of the game *Black & White* (*especially* in terms of the first and third-person undecideability) merits further investigation.

To sum up: in this *différance*-machine at no point is there a feeling of comfortable 'presence' in the game, as the state being experienced is also knowingly the state *not* being experienced. The presence of *this* game is oscillating with the (present) absence of the absent-game. The game is in play.

18
Negotiating Online Computer Games in East Asia: Manufacturing Asian MMORPGs and Marketing 'Asianness'

Dean Chan

In the past decade, online computer games have proliferated through-out the East Asian region. A key feature of this gaming context is the relationship between the distinctive use of regional aesthetic and narrative forms in game content, and the parallel growth in regionally focused marketing and distribution initiatives. Thus, intra-Asian games design and marketing play to notions of perceived cultural proximity within the region. Most Asian-designed online games are principally marketed within East Asia; and only a few, such as *Lineage* (1998) and *Ragnarok Online* (2002), have been distributed outside the region. The fact that these exported games have not been as commercially successful in North American and European territories underscores the primacy of a contextual analysis of Asian-designed games.

The East Asian online computer games boom originated in South Korea in the late 1990s. Following considerable domestic success, many South Korean games were subsequently distributed to other regional territories. Meanwhile, game companies specializing in online games for the East Asian market concurrently emerged in mainland China, Taiwan and Japan. This chapter situates topical developments in the production and distri-bution of online games in East Asia as part of broader developments in contemporary Asian regionalism, especially in relation to current models for intra-Asian cultural identification.

At the same time, however, as Chen (1998, p. 31) points out, 'when questions are asked – Is Asia one place? What *are* the Asian values? – then a universal Asian identity collapses, and differences of tradition, history and past hatreds resurface'. Yet, regionalism continues to circulate in pre-sent discourses on East Asian economies. The regional identification of 'East Asia' since the 1990s may be attributed to the global intensification of transnational capitalism. As Ching (2000, p. 257) argues: 'Asianism no longer represents the kind of transcendental otherness required to produce a practical identity and tension between the East and West ... "Asia" has become a market, and "Asianness" has become a commodity circulating globally through late capitalism.' His proposition has significant implications for an understanding of the commodity function of 'Asianness' in newly emergent intra-regional networked culture. Ching (2000, p. 249) focuses on analysing 'the discursive construction of the relationship between the concept of mass culture and regional identity'. At issue here is the question of how and to what extent 'today's Asiatic imaginary necessarily embodies different rhetorical and ideological strategies, particularly due to the global reach of capitalism in the contemporary moment' (p. 251).

Following an outline of the conditions that have enabled the rapid development of online games in East Asia, the first section in this chapter addresses the production of Asian massively multiplayer online role-playing games (MMORPGs), including an overview of culturally specific game design, narrative and game-play issues. The second section examines the marketing and circulation of the games, especially in regional contexts. Underlying and unifying the discussion in both sections is a critical analysis of the ways and means by which 'Asianness' is manufactured in Asian MMORPGs. The current industry practice of capitalizing on and marketing forms of transnational 'Asianness' is particularly evident in the popular sub-genre known as Asian Martial Arts MMORPGs.

Generally speaking, MMORPGs enable players to engage in solo and group-based interactive game-play in evolving virtual worlds. They are also known as persistent-world games in the sense that such virtual worlds are populated by thousands of other player-avatars and continue to evolve even when an individual player logs off. *Lineage* attracted significant international press attention from 2001 onwards for being the world's most heavily populated MMORPG, with more than four million subscribers worldwide (Levander, 2001). This South Korean-designed MMORPG is still regarded as one of the most popular persistent-world games, even warranting a follow-up, *Lineage II*, in 2003.

World of Warcraft, developed by the North American studio Blizzard Entertainment, has proven since its launch at the end of 2004 to be a significant global crossover success in the increasingly competitive MMORPG market. Even then, out of *World of Warcraft*'s estimated eight million subscribers worldwide at the end of 2006, three and a half million are from mainland China (Blizzard Entertainment, 2007). Networked-games culture is well and truly established in East Asia. The region is receptive to a growing range of local and imported games; however, Asian MMORPGs have been, and continue to be, instrumental in the evolution of online games networks within East Asia.

The development of online computer games in East Asia

Japan's key role in developing console-based video-games culture is unquestionable. The Nintendo Corporation was largely responsible for the global distribution and mass popularisation of NES, SNES and Nintendo 64 video games (as well as portable GameBoy games) in the late 1980s and early 1990s. Sony entered the market with PlayStation in 1994 and currently enjoys international market dominance with the PlayStation 2 console and its associated games.

Sony's PlayStation 3 performed strongly at its launch at the end of 2006 despite considerable delays and debate about its high retail price. Meanwhile, Nintendo has bounced back from the relative failure of GameCube with the runaway successes of the DS portable and Wii console launched in 2004 and 2006 respectively. This continued emphasis on developing video-game consoles and video games for domestic and international markets has arguably come at the expense of standalone PC games and online computer games in Japan.

By contrast, online games dominate in South Korea and mainland China. Many interlocking factors have contributed to the rise of online games in these territories. Console games were never officially marketed in these locations on a mass scale. South Korea, for instance, had placed restrictions on the import of Japanese popular culture following the Japanese occupation of the Second World War. These restrictions were only officially lifted in 1998. Moreover, international game companies had been reluctant to focus on the East Asian games market because of widespread software piracy. Local PC game developers in South Korea and China similarly experienced limited success in the 1990s.

In the meantime, online games culture was already evolving, especially in South Korea. Imported games such as Blizzard Entertainment's *StarCraft* (1998), a real-time strategy computer game with networked multiplayer capabilities, proved to be an early success and was a contributing factor in the mass popularization of networked computer games. *StarCraft's* iconic local status is ratified by the fact that it continues to feature regularly in televised player competitions and government-sponsored tournaments.

The rapid uptake of online games in South Korea in the late 1990s may be attributed to two interlinked infrastructural conditions, namely the expansion of national broadband networks and the proliferation of internet cafés (known in Korea as PC *baangs*). Both of these may, in turn, be linked to the Asian financial crisis of 1997–98. The governmental focus on developing the domestic information technology infrastructure as a means to rebuild the national economy, together with an attendant interest in supporting local cultural industry initiatives such as the fledgling games industry, soon produced tangible results.

By 2003, South Korea had the highest usage of broadband connections in the world. As many persistent world games rely on high-speed internet connections, the comprehensive national broadband infrastructure was undoubtedly a contributing factor in enabling the uptake of these games. Indeed, by 2003 also, South Korea had the highest proportion of online gamers per capita in the world (Chou, 2003).

Despite increasing rates of domestic computer ownership, internet cafés continue to be significant social locations for playing online games. According to the Korea Game Development and Promotion Institute, 84 per cent of internet café users play online games (KGDI, 2004, p. 22). In 2003, there were more than 20,000 internet cafés in South Korea (p. 30), where online games are played using a variety of micropayment schemes, including pay per play, hourly charges and prepaid billing cards. Such payment schemes get around the problem of software piracy and offer a measure of revenue protection for the game companies. This commercial feature has become a determining factor in fuelling the exponential growth of the online games industry within the region.

The online games development industry in South Korea continues to be supported by extensive government intervention and preferential cultural industry policies. This level of governmental backing is now replicated in mainland China, where comprehensive efforts are being made to seed the growth of the local online games development sector. For example, it was recently reported that the Chinese government is investing $242 million in the local games development industry with a view to developing more than 100 original online game titles (Feldman, 2004).

The Japanese government started to support its domestic games industry from 2001 by assisting in areas such as media content development and export-oriented initiatives. Owing to factors such as saturation of the local video-games market, declining domestic sales and Japan's persistent economic recession, Japanese companies are now increasingly concentrating on international markets and starting to expand into online games development. Perhaps the most significant example is the domestic and international distribution of Square Enix's *Final Fantasy XI* (2002), a persistent-world game that is notably part of a well-established and lucrative console-game franchise. The game is also the first cross-platform MMORPG in which PC and PlayStation 2 console versions connect to the same servers.

The official mass distribution of Japanese console-based video games in the Korean and Chinese markets in 2002 and 2003 respectively was initially successful, but market stagnation soon followed. Current schemes to improve the console-games market in these territories centre on the development of video-game network services and the introduction of networked video-gaming rooms as an equivalent to internet cafés. These scenarios are collectively indexical of the virtual hegemony of networked games and networked-gaming culture in East Asia.

Manufacturing Asian MMORPGs

South Korean-made games feature prominently in East Asian games networks. In 2002, Korean products had a 65 per cent share of Taiwan's online games market (Lin, 2002). In 2003, Korean companies controlled more than 70 per cent of the Chinese online games market (Embassy of the Republic of Korea, 2004). The three iconic Korean-designed games within intra-Asian games networks of this period are *Lineage* (1998), *Ragnarok Online* (2002) and *Legend of Mir II* (2001).

Part of the early wave of South Korean games in the late 1990s, *Lineage* relied on the then-established game-play and thematic conventions for online games. The game was modelled on European and North American paradigms for medieval fantasy role-playing games. Even then, compared to North American online gaming contemporaries such as *Ultima Online*, (1997), *Lineage* presented some cultural variations in terms of game-play design.

Firstly, there was an emphasis on in-game quests that could be completed only by highly organized groups of players (as part of teams referred to as 'blood pledges'). Secondly, the player-avatars were characterised by their allotted places within strict social hierarchies, where only members of the Prince/Princess character class can recruit groups of followers and form 'blood pledges'. These design features appear to be especially conducive to the internet café game-play context, so much so that it is not uncommon for leaders of 'blood pledges' to arrange with members to congregate in 'real life' and play together as groups in PC *baangs* (Levander, 2001).

A closer analysis of *Lineage* reveals an additional element of local acculturation. The back-story and game-world settings are derived from Il-Sook Shin's popular *manhwa* (Korean comic) of the same title. *Ragnarok*

Online shares a common point of origin. While the Ragnarok story has its origins in Norse mythology, the in-game narrative and settings are loosely based on Myung Jin Lee's *Ragnarok: Into the Abyss*, a popular *manhwa* adaptation of the Norse legend. The *manhwa*-MMORPG interface occurs at various levels. For example, the *manhwa* storyline revolves around competing guilds, thus echoing online game-play dynamics. It is also unsurprising to note that Lee was involved in the overall design of the game. Hence, *Ragnarok Online* is a persistent-world adaptation of a *manhwa* adaptation of Norse mythology; and, like *Lineage*, it is indexical of the creative intercultural and crossmedia transformations that are implicit in many South Korean MMORPGs.

Intra-Asian games networks thus partly depend on regional cross-media literacy in that the games often build on or cross-reference other popular cultural forms such as comics, animation and fantasy novels. The settings and characters in *Ragnarok Online* are very cartoon-like, especially when compared to North American game worlds, partly to reflect its *manhwa* origins, but also partly to cater to the palate for cute graphics with bright pastel colours that have become synonymous with much East Asian popular culture. The successful expansion of South Korean online games networks within East Asian markets from 2000 onwards may therefore be in part attributed to a perceived sense of cultural proximity among these territories, whereby regional cultural signifiers and themes are used as transnational markers of cultural affinity.

Legend of Mir II offers an example of cultural proximity in MMORPG design. This Korean-designed game was the most popular online game in mainland China in 2002 and 2003, attracting more than 700,000 peak concurrent users in 2003 (Actoz Soft, 2003). *Legend of Mir II* features a fantasy 'Oriental' game-world complete with a melange of traditional Asiatic design elements in architectural and dress styles, as well as a 'Taoist' character class. The overarching objective of the game is to unify and restore a once-great civilisation, thereby mining a core role-playing game trope as well as referencing a familiar narrative in classical Chinese literature. Given the commercial success of this game in mainland China and Taiwan, it seems that such generic visual and narrative design elements resonate with the present generation of Chinese-language gamers.

At the same time, however, while the term 'cultural proximity' infers notions of commoditized cultural affinity, it may invoke problematic essentialist ideas of cultural convergence, equivalence and homogeneity. At stake here is the need for a closer examination of how 'Asianness' or intra-Asian identification is modulated and marketed in Asian MMORPGs circulating within the East Asian region.

Marketing 'Asianness'

The use of marketable versions of 'traditional culture' is becoming commonplace in Asian MMORPGs. *1000 Years* (2001), for instance, is described as an 'Asian Martial Arts MMORPG' by its Korean developer Actoz Soft (2003). The promotional blurb for this game, which is simultaneously distributed in Korea, Taiwan and mainland China, reads as follows:

> Set your clock back to 100 decades ago, when the most notable historic changes occurred in the Far East. Masters of Martial Arts spread out rapidly among the three newly born dynasties of Korea, Japan and China. In this era when Kingdoms fell and new dynasties were born, players start their own journey to become a Master and rewrite the history of eastern Martial Arts (Actoz Soft, 2003).

Such visions of a shared Asian martial-arts history (however questionable) are suggestive of the manifest desire to commodify and market a sense of shared Asian cultural lineage and regional identification. As Chua (2004, p. 217) notes, 'the construction of a pan-East Asian identity is a conscious ideological project for the producers of East Asian cultural products, based on the commercial desire of capturing a larger audience and market'. *1000 Years* consistently ranked among the top five most popular online games in mainland China between 2001 and 2003 (Actoz Soft, 2003), and it is indexical of the current cultivation of Asian-specific transnational cultural networks in Asian MMORPGs.

Many new Asian MMORPGs that are aimed at regional markets seem to self-consciously invoke Asian-themed historical fantasy and martial arts. Specific cultural histories are also being represented in the process. For example, the Japanese games publisher Koei is making its MMORPG debut with *Nobunaga's Ambition Online* (2003). Tellingly, this game is based on a historical figure and set in 16th-century feudal Japan, and it features playable character classes such as 'Samurai', 'Ninja' and 'Shinto Priest'. Having already established a strong subscriber base in Japan, the game has been distributed to other East Asian territories since 2005.

Another significant example of the evocation of specific cultural histories in Asian MMORPG design is the development of Chinese *wuxia*-themed games. *Legend of Knights Online* (2003), the first major mainland Chinese-made online game, is based on *wuxia*, or martial-arts tales of knightly chivalry. Although *wuxia* stories circulated in the form of serialized novels and were incorporated into the Peking Opera in the 10th century, these fantastical and pseudo-historical tales were banned in mainland China in 1931 as part of the endeavour to create a new mass culture that would aid in the project of progressive nation-building.

Thus, from the 1930s onwards, popular culture forms based on *wuxia* were produced primarily in Taiwan and Hong Kong. Accordingly, PC games developed in these territories in the 1990s set the precedent for Martial Arts RPGs based on *wuxia* narratives (Liu, 2001). Many of these were based on the popular martial-arts novels of Louis Cha (a.k.a. Jin Yong), and with the present turn to online games in the region, Taiwan in particular has continued to develop games such as *Jin Yong Online* (2001), mainly for domestic and regional Chinese-language audiences.

The dominant form of Asian Martial Arts MMORPGs is the *wuxia*-styled persistent-world game. According to one report, *wuxia* games constitute one third of the online games market in China today ('China busy', 2004). As one games publisher notes: 'The emerging strength of Chinese Wuxia-style [...] online games demonstrates that Chinese gamers are hoping to see their own traditional values and specific historical artifacts in the new cyber-realities' (Kim, 2004b).

However, what kind of 'tradition' is being engaged here? Liu Shifa, a spokesperson for China's Ministry of Culture, asserts, '[*Legend of Knights Online*] proves the charm of homemade online games, which have begun to serve as a catalyst for the rebirth of the whole information industry' (cited in Xinhua News Agency, 2003). On one level, *Legend of Knights Online* may be regarded as 'a hybrid that engages with the tradition of the *wuxia* genre and with the process of cultural production at a specific historical moment in shaping a cultural identity' (Lee, 2003, p. 281). Yet, on another level, the game underscores the process whereby *wuxia* narratives are now proactively recuperated in China as a sign of indigeneity and fashioned into a marketable aesthetic.

Asian antiquity (imagined or otherwise) acts as a common reference point for in-game narratives, characters and imagery in many Asian MMORPGs. These games stage a performative articulation of legible difference, especially in their capacity to act as visibly different and localized cultural products that may be distinguished from other global cultural products. In this respect, Asian Martial Arts MMORPGs offer an alter/native counterpoint to Euro-American fantasy paradigms for persistent-world games. Asian antiquity acts as a trope of authentication and difference. However, this citation of antiquity is not always narrowly nostalgic or nativistic in orientation because 'authenticity' is used as a means to distinguish locally produced games without necessarily disavowing the significance of imported forms and borrowed styles.

These machinations are underscored in Actoz Soft's (2003) promotional blurb for *Legend of Mir II*: 'While most RPGs are focused on North European Fantasies, *Legend of Mir II* strongly emphasizes [...] an original story with oriental background, mixed with western type RPG elements'. In this sense, Asian MMORPGs may be regarded as modern popular-cultural forms that are simultaneously marked as local and international, as specifically 'Asian' but always already hybridized in orientation. At any rate, cultural hybridization does not necessarily result in the levelling of differences. Differing registers of 'Asianness' are taken into account in the regional distribution of online games, so much so that the same game may be played and experienced somewhat differently in each territory.

Intra-Asian games networks are sustained by the standard East Asian game-development practice of providing customizable territory-specific content and extensive localization services for products that are distributed regionally. According to Jung Ryal Kim (2004a), the chairman of Gravity Corporation, the South Korean developer of games such as *Ragnarok Online*, there is a twinned process involved in intra-Asian games distribution – namely, localizing in-game content and gameplay mechanics to make the game familiar to target users, and using local hosting partners to assist in the ongoing provision of game services.

Regional localization processes are thereby contingent on the establishment of collaborative transnational ventures within intra-Asian games networks. For instance, the Japanese games publisher Square Enix entered into partnership with Webstar (an affiliate company of Softstar Entertainment, Taiwan) for its first foray into the online games market in mainland China in 2002 with *Cross Gate* (2001), a MMORPG developed specifically for the Asian market. At issue here is the significance of localized cultural knowledge.

As Ragaini (2004) notes: '[T]here's a tendency to oversimplify the significant regional differences between the various countries. Singapore, Indonesia, Japan, China, and South Korea should all be considered separate marketplaces with distinct needs, expectations and system specifications.' At the most basic level, localization requires both the straightforward linguistic translation of the game text and the provision of territory-specific content. For example, Kim (2004a) points out that Gravity utilizes 'region-specific updates that allow players to enjoy replicas of historical buildings, wear traditional indigenous apparel, fight creatures inspired by local myths, and collect culturally themed items'. Moreover, in *Ragnarok Online*, players may visit and congregate in different virtual cities designed in ancient Korean, Japanese, Taiwanese, Chinese and Thai styles. It is no coincidence that this sample is reflective of the main markets for this game.

Preferred game-play styles are also different in each territory. Eds. Tan (2004), the CEO of Phoenix Games Studio, provides the rationale for territory-specific variations in *Fung Wan Online* (2003), a martial arts persistent-world game based on a popular Hong Kong comic. For example, a pickpocketing feature was disabled in Taiwan, but left intact in Southeast Asia. Tan's research found that although both markets enjoy player-versus-player combat, 'the magnitudes of punishments and rewards' are nonetheless considerably different.

Such localized interventions ensure a degree of cultural familiarity and relevance in different territories. The processes of regional localization therefore provide insights into the intricate modulation of 'Asianness' within intra-Asian games networks. 'Asianness' is crucially not treated as a singular and unchanging referent. Instead, the plurality of Asian audiences is tacitly underscored in intra-Asian games localization.

The dynamics of transnational localization initiatives and joint-venture operations are quickly evolving, underpinning the creation of increasingly more expansive games production and distribution networks throughout Asia, particularly between East and Southeast Asian territories. For example, in 2005, the Japanese games publisher Koei opened up a new software-development centre in Singapore, its first one outside Japan. While the new subsidiary will be initially concerned with localization, there are plans for it to subsequently design, develop and market its own original titles. Keiko Erikawa, Koei's chief executive, is hoping that Singaporeans can create games with a market-ready Asian 'flavor' (Fraioli, 2003).

Koei Singapore's first project is to develop an online version of the publisher's popular turn-based strategy video game *Romance of the Three Kingdoms*, which is based on the classic Chinese historical novel with the same title. The online game is scheduled for release in 2007 in countries across the region including Singapore, Japan, mainland China, Taiwan and Korea (Yu, 2005). Indeed, one might ask, how will such emergent transnational arrangements, together with current plans to distribute more Asian-designed games beyond regional territories, affect the content and design of these online games in the near future? Suffice to say at this stage that intra-Asian games networks offer a rapidly evolving context for continued study and further critical examination.

Conclusion

The contextual analysis of Asian-designed persistent-world games in this chapter draws attention to the complexities inherent in transnational East Asian cultural production, regional cultural flows and intra-Asian identification, especially in terms of how 'Asianness' is subject to intricate processes of modulation, localization and hybridization within intra-Asian games networks. While South Korean games continue to feature prominently within these networks, increasingly proactive online games-development initiatives are emerging in mainland China, Taiwan and Japan. The current regionally focused marketing of Asian MMORPGs – and Asian Martial Arts MMORPGs in particular – is perhaps unsurprising, given that a certain degree of cultural familiarity is required for fully appreciating and engaging with the sociocultural contexts inscribed in the game-world settings. As the types and content of East Asian online games will undoubtedly continue to evolve and diversify, this is therefore a timely moment to critically review the significant contributions made by Asian MMORPGs in cultivating contemporary intra-Asian games networks.

19
Teenage Girls 'Play House': The Cyber-drama of *The Sims*

Lynda Dyson

The Sims has been described as a 'software toy' or an 'electronic tamagotchi'. The game provides an 'authoring environment' that enables players to create digital 'doll-houses' inhabited by simulated human-type avatars capable of interacting and responding to their environment (Jenkins, 2004, p. 128). Importantly for the girls in this study, game-play is open-ended – there is no means 'to win'.

The identity of each avatar is created by selecting from a range of personality and appearance traits which produce an identikit character constructed in terms of gender, skin colour, age, appearance and 'sensibility' (neat, outgoing, playful, friendly, active or nice). It is possible to create *Sims* 'skins' which resemble 'real-life' people, albeit in cartoonish form. These characters inhabit suburban-style household spaces designed by the player. The process of constructing a household – buying houses and furnishing them, giving the *Sims* character a distinctive style – makes consumption an important aspect of *The Sims*' life-world. For example, a way to increase characters' happiness and aspiration quotients (and thus to keep them alive) is to 'buy' them things from the *Sims* catalogue. The artefacts purchased are used to augment the environment of the characters by performing specific functions – newspapers carry job information, bookcases make characters more intelligent, love beds make babies while cheap beds require more sleep-time.

Within the domestic space of the suburb *Sims* characters ('Sims') can interact with other Sims through their assigned roles as fathers, mothers, workers, lovers, children, neighbours and friends. Characters are given desires, urges and needs and respond emotionally to events, coming into conflict with one another to produce dramatically compelling encounters. Players generate social and domestic scenarios that have no necessary narrative resolution but are focused primarily on interpersonal romance and conflict, while at the same time coping with the fairly mundane processes of social reproduction – cooking, taking showers, sleeping and so on.

The experience of playing is 'god-like' in as much as the player's point of view over the game terrain resembles that of a movie camera suspended over a set. The game produces a 'first-person perspective', with the player suspended over the navigable space, controlling the point of view and directing the spatial position of the Sims, thus authoring social interactions whilst not necessary controlling outcomes.

This article is based on interviews with a group of four teenage *Sims* players who have played the game regularly over a number of years. Throughout the study there have been methodological difficulties involved in getting the girls to talk about their relationship to the game. The girls were interviewed, alone and in pairs, as they played the game. They often found the interviews disruptive and intrusive and sometimes only consented because it was financially beneficial to do so – they were paid £10 for each interview session.

Although the girls who took part in this research were not necessarily able to articulate the reasons why the game has been so important in their lives (to them 'it's just fun'), it is clear *The Sims* provides a legitimate space for play in an urban context where peer-to-peer- and self-policing closes down the possibility of such 'childish' activities in public domains. As one of the respondents, Erica, told me when she was 14: '*Sims* is much better than talking to real people! It's all your own ideas. It's so personal to you.'

The game has been important to these girls throughout their adolescence. The four, aged between 15 and 16 at the time of writing, continue to play, albeit in very different ways from when they were 12. Like so many of their generation, they live much of their social lives through the screens of mobile phones and computers. Their favourite modes of communication are mobile-phone texting, the instant-message facility MSN and its links to personal web domains through MySpace. These technologies enable them to remain electronically connected to their friends at all times although outside of school they spend much of their time in physical isolation from one another.

For numerous reasons their social lives in London have, throughout their adolescence, been relatively proscribed. Hackney is a deprived inner-city borough with increasing numbers of middle-class, property-owning enclaves. There are highly publicised levels of street crime and the girls' parents have been protective, restricting their movement outside the home (particularly in the evenings). Financial constraints also prevent extensive socialising outside the home and, most importantly for their game-playing, the girls are highly attuned to the ever-present threat of peer-to-peer bullying and harassment on the street, a threat which oppresses all of them despite the fact that none of them has actually been physically bullied ('We bully ourselves,' one of them told me during an interview).

For each of the girls the game offers a highly individuated and private space for play and is used in different ways at different times, to create utopian and dystopian fantasy realms. Depending on what is going on in their lives at any time, they may use the game to create households and relationships mimicking certain aspects of their experience or they may focus on the spatial possibilities offered by *The Sims* – constructing living spaces, inventing architecturally interesting houses, and developing distinctive designs which reflect and elaborate certain taste cultures in their 'real-life' milieu.

The game itself is expensive to both purchase and, perhaps more crucially, because it needs to be played on a powerful PC (*Sims 2* requires enhanced graphics capability and thus significant computer memory). As a result, two of the girls in the study (Jade and Erica) have always played the game with other people. Inevitably these girls came from poorer households with no domestic computing facilities. These two have been brought up by single parents – Jade by her white English father and Erica by her second generation Afro-Caribbean mother. The other two – Libby, who describes herself as 'mixed-race', and Lisa, who is white – live in families with step-parents and siblings and use the game as a means of escaping the pressures of these domestic arrangements.

The cybergeography of *The Sims* bears no relation to the living spaces of the inner city. The navigable suburban space of the game stretches like a smooth green lawn to the edge of the computer screen and references, for the London girls, the privileged settings of US television series and films rather than the deprived urban landscape they inhabit. During interviews they have frequently spoken about the influence of North American cultural norms on the design of the game and have remarked on the way their language, when playing, takes on the argot of American popular culture: for example, 'making out' is a favourite term for early forms of sexual relating between *Sims* characters.

One of the primary pleasures of the game for these girls has been the capacity to reconfigure their lives in a fantasy space created and manipulated to suit themselves – it is one of the few spaces they can make their own. They have spoken of the pleasures involved in reproducing versions of themselves inside *The Sims*. This performative aspect of the game – the way the player has the capacity to construct individuated characters and scenarios – makes *The Sims* a fascinating object of study, but also creates methodological difficulties for an 'ethnographic' approach to game-play.

Perhaps unsurprisingly, during the course of interviews the girls tended not to discuss their *Sims*-play in terms of their own personal lives; they often expressed concern about how the interview material was going to be used. However, they frequently spoke about the way they used the game to act out scenarios between characters based on people they knew. When interviewed in pairs, they would often point out to each other how their game-play was self-reflexive. As one of them put it: 'You live your life through the game. You project so much of yourself onto the characters and that's why it takes you over.'

Methodologically, the contextual lacunae (the way the girls did not connect their 'real-life' situations with their game-play in any explicit way) presented ethical difficulties in relation to the interpretation of interview material. The temptation to 'read off' certain aspects of game play in relation to information gleaned about the players' background and circumstances invited an overly 'psychologised' form of interpretation that had to be resisted. It also seemed important to resist interpretation of their attitudes to race and sexuality despite the fact that these were important aspects of their *Sims* identities.

And of course, given the girls were secretive about the most private aspects of their game-play, the interview material inevitably generated fairly limited accounts of the intimate and complex ways in which 'play' has actually taken place. They would sometimes retrospectively 'own up' to instances where they had kept certain details to themselves once it felt safe to do so. One confided: 'I remember when I had just discovered how to get my characters to use the love bed but I didn't tell you.'

However, the most difficult aspect of this approach to studying the game has been the disjuncture between the way in which the players narrate the details of game-play to a third party (in this case the interviewer) and the way in which they themselves are engaged with their characters and the scenarios they create. There is no fixed 'text' to refer back to – the interview material refers to characters and games that are in a state of dynamic transformation for a time and tend to eventually disappear because the girls do not archive the characters or the architectural spaces. In fact the girls themselves have often not been able to remember households and scenarios they created in the past.

One of the pleasures of the game seemed to be the very impenetrability of scenarios on the screen to non-players. *Sims* relationships and situations need to be explained or narrated to outsiders and this, of course, enhances the possibility of secrecy and privacy. Play is contingent and ahistorical. Scenes appear and disappear like mirages and, unlike other visual texts (such as films or television programmes), there is no possibility of 'freezing' or recording the flow of the game. There is no real text here, just so many instantiations – the material is almost dreamlike, subjective and autobiographical. (Despite the possibility for players to create *Sim* 'memories' by taking photographs and creating albums, none of the girls in this study used this facility.)

In the main, social scenarios seem to hold the most interest for the teenagers. Throughout the interviews they frequently spoke of the pleasures involved in playing the game on their own. As one of them put it: 'I love going into my own world with my *Sims* characters. It's better than talking to real people!'

In the following interview extract, Jade describes a series of relationships she has created in a neighbourhood. At the time of this interview she was 14 and had been using the game to explore the troubling gender relations she was experiencing in her own life:

> Jonathan, Sarah and Amy live there (points to house on screen). Jonathan started seeing Mrs Wilkins next door. She knocked on the door, he asked her in and they had drinks. She came from that house (points to another house on screen). Sarah and Amy keep having arguments and slapping each other. Jonathan keeps cheating on both of them ... he gets all the benefits and they feel really down about it sometimes and Jonathan and Sarah had a baby but Amy's always looking after their son. Sarah's having an affair with Daniel I think his name is and when the Mum goes out they have sex.
>
> I think I got the idea from films and some boys' personalities at school. The way they talk about how when they're older they're going to be really macho. I kind of used it in *Sims* to see how it would look like, how it would be. But there's not as much drama as I thought there would be ... What happens is, the men hardly ever get hit, it's normally the woman who hit each other. Jonathan never gets hit, it's only the two girls who slap each other – I don't know why that is ...

Here a stereotype of heterosexual relationships appears to be reproduced in Jade's convoluted scenario where, puzzlingly for her, the promiscuous male character she has created produces jealousy and resentment between the two female characters who engage in catfights while he manages to escape retribution himself. During the interviews the girls did not recount a single instance where two men fought over a woman (or another man) but this may be simply because they tended to create certain 'romantic' scenarios in their game-playing which produced the same kind of jealous reactions in their Sims.

As a mass-marketed media object, *The Sims* initially appears 'normative' but part of its appeal to the teenagers is the way the game can be used for subversion and transgression, a secret aspect which is easily kept from the adults attempting to police the teenagers' behaviour by censoring their consumption of cultural products.

In order to learn the game, players first encounter a 'ready-made' sample, the household of a heterosexual couple, Bob and Betty Newbie, who live in a suburban house with an array of objects appropriate to their 'lifestyle' – televisions, computers, refrigerators and so on. Libby admitted she liked this sample family because it was so 'normal'. She had domesticated the Newbies by incorporating the family into her game. Interestingly she embellished the household with a female character who reminded her of a local girl:

> This is a house I really like, it was already in *The Sims*, the Newbies. It's the kind of house which is almost a perfect family except for the girl. It's what a lot of people would call a perfect family. The parents get on, there's been no affairs, kids get on with each other mostly and the adults and kids get on. The girl's called Epiphany ... I love that name ... it's from a girl who's quite well known round here who's quite rough. She beats a lot of people up with baseball bats. She feels bad because she has to wear glasses and her hair's stuck in a position where she has to wear bunches. I can't change it for her.

Given the teenagers are trying to work out who they are and what they are to become, identity issues, in terms of gender, 'race', sexuality and class, haunt their game-playing.

The girls in this study always use the three skin shades available to construct 'racialised' characters. Skin tone is never neutral in their world because 'race' is an everpresent issue in inner London. The secondary school the girls attend is, like most London schools, divided along race and class lines. Here is Jade (who is white) discussing her experiments with different *Sims* 'skins':

> That's the gangsta family's house (points to house). They're all mixed-race, with one black person ... that's partly from having kids. I didn't

really think about racism that much, I just made these people but what I did find was that this happened: if you want to make people have kids they have to be the same colour to make it work ... Apart from the mixed-race and black people ... you can have black and white ... if there's a white woman they seem to not get on so well ... you have to get them to like each other to get them into bed ... you've got to build it up ... what happens ... they don't seem to get on ...

Lisa describes herself as 'mixed race' and has often spoken of making characters that resemble her. Her playing has become focused on ensuring the aspirations of her *Sims* characters are fully realized and she is able to articulate how their success reflects back on her:

I used to make trailer-trash families but now I don't cos I don't get fun out of it – it's boring. In my games women are dominant and their skills are maxed out. When they achieve their aspirations I think, yes, I'm brilliant! The game can get really nasty and destructive and I don't like that ... it gets me really annoyed if the Sims have a low aspiration ... if I want to get maxed-up aspiration and them to have a happy life I don't want all these bad things going on because it affects their aspiration meter.

It became clear from the interviews that as the girls familiarised themselves with the parameters of the game, the possibility of pushing the boundaries by engaging in 'transgressive' activities offered particular pleasures. The degree and content of the 'transgressions' were age-related and linked to their social and sexual stages of development. In the early era of play, as 12- and 13-year-olds, the girls began to experiment with 'baby-making'; they then moved on to creating same-sex relationships; much later they started 'murdering' their characters.

Here is Erica in the earliest days of the study, describing what happened when she bought a bed for 'parents' who, in her *Sims* world, are the only avatars having sex:

For the parents' bedroom – you can get different beds ... This bed you can vibrate on ... I've never used another bed before to try and make two parents have a baby ... what happens is you get one parent in the bed and make it vibrate then you click on the other parent and then you press on the bed and it says 'play in bed'. You see them rummaging and then you hear 'woof woof' and then a question pops up 'do you want a baby' then a daisy appears and then a baby comes down and you have to look after it for three days and if you don't look after it it gets taken away by social services.

Despite the excitement generated by the possibility of 'making out', the babies that can be the product of this activity (if the players so choose) have proved very unpopular with the girls. In the extract below, two of them discuss the problem of *Sim* babies. The discussion is typical in the way the subjectivity of the players is entwined with the subjectivity of their *Sims* characters:

> Lisa: I like having babies!
> Libby: I hate it!
> Lisa: It's true, they become a nuisance; they're so much work!
> Libby: You start with a baby and then you have to get a nanny, otherwise social services comes around and takes it away.
> Lisa: You have to feed them all the time.
> Libby: If you have a baby it gets the characters down, have you noticed that? The parents don't achieve any aspiration wants ... their happiness and energy decreases. They take over your life, it's hard. When they're pregnant they can collapse because they're eating for two. Sometimes they can die!
> Lisa: They can't have sex while they're pregnant. When they have a baby the whole house gets turned upside down. When you see people playing the game you can get a bit of an idea about what they're like. I bet you don't want to have a baby! You want a career!

In another interview, 12 months later, Lisa talks of trying to 'kill' a baby who was ruining a couple's aspirations. Her ironic reflection on the way she speaks about her characters gives a hint at the way game-playing provides an involvement which is never totalising – the girls move in and out of the *Sims* 'reality':

> That couple were at the top of their careers so it was natural for them to have a baby (laughs). What am I like? I don't have weird views, you know! But I didn't like the baby so I tried to kill it by removing objects ... but somehow it didn't work and the crying was so realistic it made me feel guilty. The crying of the babies is so terrible. That's the thing, it's so realistic ... the interactions between people. Social services came and took that baby away

At various times all the girls have taken pleasure in their attempts to torture and kill their characters. They would often recount these episodes by first explaining that someone else had taught them how to do it. Libby claimed Jade was the first in the group to discover 'torture': 'She used to stop feeding them then take the walls down in her houses and watch them slowly die!'

Libby herself has dispensed with characters:

> I had one woman and she moved her boyfriend in but I didn't like him, not because he was a firefighter but because his aspiration was 'romance' and that didn't suit her so I got rid of him. Then I created someone else for her but he annoyed me .... I didn't give them a smoke alarm and he got killed because he hadn't learnt to cook. I was pleased about that.

When she was 13 Erica shyly confessed to making a lesbian household:

> These characters are both mixed-race. They're called Caira and Maira. I wanted to test it out ... to see if two of the same sex would have a relationship and get into bed. I made them talk and then give each other gifts then hug and then kiss and it worked ... they got into bed. A baby option did come up but I didn't think they should have kids. I think they are quite nice people though.

When the girls make *Sims* characters based on people in their own lives it is apparent that the game provides a way of fantasizing about peers who can be simultaneously frightening and attractive. In the account below, Lisa, who almost always uses 'cheat' codes to provide her households with luxury items and a swimming pool, talks about creating a house inhabited by a 'bad boy' she knows from her locality.

Unusually, she does not use the 'cheat' codes to create his household, but only the money allocated by the game. In the interview she emphasizes that in doing so she was not being punitive to the particular characters, but that it was just an experiment:

> This family is quite interesting. It reminds me of Chantelle and Sirus Jordan. There's a boy who's well known in this area ... his name's Sirus, he's got into trouble with the police and a lot of people know about him ... a girl called Chantelle who goes to my school, she used to fancy him. I decided to put these two together ... I made him not very nice cos he's not very nice anyway ... They've got a house where I just used the amount of money you can have with the game – they've got a cheap chair, cheap desk, cheap computer ... cheap toilet ... I didn't

punish them ... I just wanted to test a family out to see if they could have a nice life with the money *The Sims* provided them with.

Conclusion

The teenage girls in this study inhabit a social space where cultural activities are increasingly individuated and privatised. There is a sense in which the panoptical perspective offered to players of *The Sims* replicates the control teenagers experience in their day-to-day lives. Teenagers are policed at school and in the home; they tend to be demonised in public urban spaces (there are endless moral panics about their generation's alienation from the aspirations of the mainstream) and while railing against this surveillance they also paradoxically subject one another to high levels of control over appearance and behaviour.

The narrative possibilities offered by *The Sims* through avatars and simulated objects affords players an arena for fantasy play in a social and cultural milieu where peer-to-peer self-policing closes down playful interactions. The fact that they live much of their lives through screens could be characterised as a 'retreat' from the public sphere engendered by commodification and individualization. However, the interview material gathered for this study shows that *The Sims* is a game which affords players an opportunity for performative enactments which have the possibility to be imaginative, creative, transgressive, banal, normative and so on.

In other words, by creatively engaging with *The Sims* the consumer/player becomes a 'producer'. The god-like position of the player subjects the Sims to forms of surveillance and policing that mimic the way teenagers experience their day-to-day lives. By inventing and controlling the game they have managed to take control; the navigable space of the game becomes a subjective space, acting like a mirror held up for the user.

Cited Computer Games

1000 Years (2001), Actoz Soft, 192
America's Army (2002), US Army, 16, 72, 84, 86, 93–95, 108, 168, 172
Animal Crossing (2002), Nintendo, 32
Antiwargame (2003), Futurefarmers, 85
Asheron's Call (1999), Microsoft, 79
Asteroids (1981), Atari, 6

Banjo-Kazooie (1998), Nintendo, 49
Black & White (2004), Electronic Arts, xvii, 127, 183, 185
Breakout (1976), Atari, vi, 162, 164, 165, 168, 170–2
Brothers in Arms: Road to Hill 30 (2005), Ubisoft Entertainment, 17

City of Heroes (2004), NCSoft., vi, xiii, 140–9
Civilization (series), Microprose, 10, 12, 32, 119
Civilization II (1996), MicroProse Software/Activision, 32, 119
Civilization III (2001), Firaxis Games, 32
Codename: Panzers (2004), CDV, 134, 138
Command & Conquer [C&C] (series), Westwood Studios, 10, 11, 78, 137
Command & Conquer: Red Alert 2 (2000), EA, 78
Command & Conquer: Yuri's Revenge (2001), EA [expansion pack], 78
Commando Lybia (1986), Activision/ Sega, 157
Commandos: Behind Enemy Lines (1998), Eidos Interactive, 138
Counter-Strike (2000), Vivendi Universal, 11, 71, 82, 169
Cross Gate (2001), Enix, 194

Dark Age of Camelot (2001), Mythic Entertainment, xiii
Day of Defeat (2003), Activision, 49
Dead To Rights (2002), Namco, 9
Death Race (1976), Exidy, 156
Defender (1980), Williams Electronics, 21, 39, 80, 81, 83, 85, 86
Deus Ex (2000), Eidos Interactive, 132, 133, 137, 138
Diablo (1996), Blizzard Entertainment, 49, 83, 138
Diablo II (2000), Blizzard Entertainment, 49, 83
Doom (1993), id Software, xiv, 42, 56, 69–71, 73, 92, 102, 135, 138, 157, 166, 167, 181
Doom II: Hell on Earth (1994), GT Interactive, 92
Duke Nukem 3D (1996), Apogee Games, 71

Enter the Matrix (2003), Shiny Entertainment., 12, 137
Escape from Woomera (2003), Escape from Woomera Collective, 18, 168
Eternal Darkness: Sanity's Requiem (2002), Nintendo, 43, 44, 53, 54
Ethnic Cleansing (2002) Resistance Records, 151

EverQuest [EQ] (1999), 989 Studios, xvii, 27, 56–65, 79
Eyewitness (unpublished), Hong Kong Polytechnic Institute, 19

Fable (2004), Microsoft, 127
Fahrenheit (2005), Sony, 26, 72
Fallout: A Post Nuclear Role Playing Game (1997), Interplay, 135, 138
Far Cry (2004), Ubisoft, 135, 138
Final Fantasy (series), Square Enix, 7, 189
Final Fantasy XI (2002), Square Enix, 189
Frogger (1981), Sega, 100
Full Spectrum Warrior (2004), THQ, 72, 84, 86, 108
Fung Wan Online (2003), CiB Net Station, 195

Gothic (series), Egmont Interactive, xi, 136, 138, 144
Gran Turismo 3 (2001), Sony Computer Entertainment, 48
Grand Theft Auto series, Rockstar Games, 6, 32, 41, 49, 55, 72, 152
Grand Theft Auto: San Andreas/ GTA San Andreas (2004), Rockstar Games, vi, xviii, 133, 138, 150–61
Gungrave: Overdose (2004), Mastiff, 9

Half-Life (1998), Sierra Entertainment, 32, 35, 36, 42, 52, 83, 133, 135, 138
Hitman: Contracts (2004), Eidos Interactive, 108, 111, 112
Hunt the Wumpus (1976), Gregory Yob, 53, 55
Hunter, In Darkness (1999), Andrew Plotkin, 53, 55

Ico (2001), Sony, v, x, 16, 26–8, 30, 31, 37, 38, 51, 53, 56, 57, 59, 61, 63, 65, 101, 148, 172, 188, 190

JFK Reloaded (2004), Traffic, 8, 11, 14, 21
Jin Yong Online (2001), Chinese Gamer International Corp., 193

Katamari Damacy (2004), Namco, 100
Kuma/war: The War on Terror (2005), Kuma Reality Games, 86

Legend of Knights Online (2003), Kingsoft, 192, 193
Legend of Mir II (2001), Shanda Interactive Entertainment, 190, 191, 194
Legend of Zelda: The Wind Waker (2003), Nintendo, 49, 53
Lineage (1998), NCsoft, 43, 44, 186, 187, 190–2
Lineage II: The Chaotic Chronicle (2003), NCsoft, 43, 44, 186, 187, 190–2

Madden NFL (2002), EA Sports, 20, 83
Mafia: The City of Lost Heaven (2002), Gathering of Developers, 111, 138
Majestic (2001), Electronic Arts, 34, 78
Manhunt (2003), Rockstar Games, 6, 75
Maniac Mansion (1987), Lucas Arts, 153
Marathon Infinity (1996), Bungie Software, 164

Max Payne (2001), Rockstar Games, 9, 133, 137
Medal of Honor: Rising Sun (2003), Electronic Arts, 12
Mercury (2005), Ignition Entertainment, 100
Microsoft Flight Simulator (2002), 71, 78, 79

Need for Speed: Most Wanted (2005), Electronic Arts, 138
Nobunaga's Ambition Online (2003), Koei, 192

Operation Flashpoint (2001), Codemasters, 80–3, 134

Pac-Man (1979), Midway, 162
Pong (1972), Atari, xvi, 39, 162
Prince of Persia: Sands of Time (2003), Ubisoft, 8, 9
Project Gotham Racing (2001), Microsoft Game Studios, 48, 54

Quake (1996), id Software, xiv, 56, 57, 73, 164, 167

Ragnarok Online (2002), Gravity Corporation, 186, 190, 191, 194
Rainbow Six: Rogue Spear (1999), Red Storm Entertainment, 80–2, 84
Real War (2001), Simon and Schuster, xvii, 72
Resident Evil (1996), Virgin Interactive, 116
Return to Castle Wolfenstein (1983), 135, 138
Rez (2001), Sega, 34, 40, 52, 55
Romance of the Three Kingdoms (1985–2007), Koei, 195

Sega GT 2002 (2002), Sega, 48
Shinobi (1987), Sega, 8
Silent Hill (1999), Konami, 33, 51
Sim City (1989), Maxis, 106, 135
Sims, The (2000), Electronic Arts, vii, xix, 32, 34, 36, 83, 137, 197–206
Space Invaders (1978), Taito/Midway Games, 81
Spacewar (1962), Steve Russell, 69, 70
Stalker (forthcoming), THQ, 135, 138
Star Wars: Knights of the Old Republic (2003), LucasArts, xviii, 42
StarCraft (1998), Blizzard Entertainment, 188

Tetris (ca. 1986), Alexy Pajnitov, 39, 100, 104, 106
The Fall: The Last Days of Gaia (2004), Deep Silver, 135
Tomb Raider (1996), Eidos Interactive, 164

Under Ash (2002), Dar al-Fikr, 86
Under Ash 2: Under Siege (2005), Dar al-Fikr, 18, 86, 133
Unreal (1998), GT Interactive, 19, 21, 84, 85, 164, 165, 170
Urban Chaos (1999), Eidos Interactive, 153

Vampire – The Masquerade: Bloodlines (2004), Activision, 119, 137
Vampire – The Masquerade: Redemption (2002), Activision, 119

Waco Resurrection (2003), C-Level, 13, 17, 19, 168
Warcraft series, Blizzard Entertainment, xiii, xvii, 7, 114, 134, 138, 187
Warcraft III (2002/03), Blizzard Entertainment, 134

Zaxxon (1983), Sega, 6
Zork (1980), Personal Software/Infocom, 53, 140

References

9/11 Survivor Website (2003), http://selectparks.net/911survivor/911about.html [Accessed 10 February 2007].

Aarseth, Espen (1997) *Cybertext: Perspectives on Ergodic Literature* (Baltimore: Johns Hopkins University Press).

Actoz Soft (2003) 'Real Entertainment Real Difference', http://www.agdc.com.au/03presentations/phpslideshow.php?directory=actoz_soft [Accessed 25 August 2005].

Adams, Ernest and Andrew Rollings (2003) *On Game Design* (Indianapolis: New Riders).

Afkar Games (n.d.) 'English Download Page', *Afkar Games*, http://www.underash.net/en_download.htm [Accessed: 8 December 2005].

Akira (2003) 'History of @war', *Atwar: Assault and Tactical Warfare website*, 18 September, http://www.atwar.net/content.php?article.12 [Accessed 20 September 2004].

Allen, Vernon L. and David B. Greenberger (1978) 'An Aesthetic Theory of Vandalism', *Crime and Delinquency* 24 (3), pp. 309–21.

Althusser, Louis (2001) 'Ideology and Ideological State Apparatuses', in *Lenin and Philosophy and other Essays*, trans. Ben Brewster (New York: Monthly Review Press).

America's Army FAQs (n.d.), United States Army, http://www.americasarmy.com/support/faq_win.php [Accessed 15 September 2005].

Anderson, Craig A. and Karcu E. Dill (2001) 'Video Games and Aggressive Thoughts, Feelings and Behaviour in the Laboratory and in Life', *Journal of Personality and Social Psychology* 78(4), pp. 772–90.

Anderson, Craig A. (2004) 'An Update on the Effects of Playing Violent Video Games', *Journal of Adolescence* 27 (1), pp. 113–22.

Angelfire Website (n.d.), http://www.angelfire.com/games4/wtc/.

Bakhtin, Mikhail (1984) *Rabelais and His World* (Bloomington: Indiana University Press).

Bandura, Albert (2001) 'Social Cognitive Theory of Mass Communication', *Media Psychology* 3 (3), pp. 265–99.

Bandura, Albert (2002) 'Selective Moral Disengagement in the Exercise of Moral Agency', *Journal of Moral Education* 31 (2), pp. 101–19.

Barthes, Roland (1975) *The Pleasure of the Text* (New York: Hill and Wang).

Baudry, Jean-Louis (1985) 'Ideological Effects of the Basic Cinematic Apparatus', in Bill Nichols (ed.) *Movies and Methods, Volume II* (Berkeley: University of California Press).

Bauman, Zygmunt (1993) *Postmodern Ethic* (Oxford: Blackwell).

Baumgärtel, Tilman (2004) 'Zu einigen Themen künstlerischer Computerspiele', http://www.medienkunstnetz.de/themen/generative_tools/game_art/ [Accessed 10 October 2005].

BBC News (2001a) 'Game Withdrawn After Attacks', *BBC News*, 13 September, http://news.bbc.co.uk/1/hi/entertainment/new_media/1543112.stm [Accessed 10 January 2007].

BBC News (2001b) 'Gaming Industry to Review Content', BBC News, 14 September, http://news.bbc.co.uk/1/hi/entertainment/new_media/1544813.stm [Accessed 10 January 2007].

Behind the Scenes (2004) *Behind the Scenes: Making of Halo 2*, Dir. James McQuillan.

Benjamin, Walter (1969) *Illuminations: Essays and Reflections*, ed. Hannah Arendt, trans. Harry Zohn (New York: Schocken Books).

Bennington, Geoffrey and Jacques Derrida (1993) *Jacques Derrida* (University of Chicago Press).

Bennington, Geoffrey (n.d.) '*Seulemonde* Conversation with Geoffrey Bennington', *Seulemonde*, http://www.cas.usf.edu/journal/bennington/gbennington.html [Accessed 3 January 2007].

Berger, Arthur A. (2002) *Video Games: A Popular Culture Phenomenon* (New Brunswick, NJ: Transaction Publishers).

Berlant, Lauren and Michael Warner (1999) 'Sex in Public', in Simon During (ed.), *The Cultural Studies Reader* (London; New York: Routledge).

Bersani, Leo (1987) 'Is the Rectum a Grave?', in Douglas Crimp (ed.) *AIDS: Cultural Analysis/Cultural Activism* (Cambridge, MA: MIT Press), pp. 197–222.

Blast Theory (2001) 'Can You See Me Now?', http://www.blasttheory.co.uk/bt/work_cysmn.html [Accessed January 2007].

Blizzard Entertainment (2007) [Press Release] '*World of Warcraft* Surpasses 8 Million Subscribers Worldwide', 11 January, http://www.blizzard.com/press/070111.shtml [Accessed 22 January 2007].

Boal, Augusto (1985) *Theatre of the Oppressed* (New York: Theatre Communications Group).

Bogost, Ian (2005) 'Frame and Metaphor in Political Games', in Suzanne de Castell and Jennifer Jenson (eds) *Changing Views: Worlds in Play, Selected Papers of the 2005 Videogames Research Association's Second International Conference* (Vancouver, BC: Videogames Research Association) pp. 59–68.

Bogost, Ian (2006) *Unit Operations: An Approach to Videogame Criticism* (Cambridge, MA: MIT Press).

Bogost, Ian (2007) *Persuasive Games: The Expressive Power of Videogames* (Cambridge, MA: MIT Press).

Bolter, Jay D. and Richard Grusin (1999) *Remediation: Understanding New Media* (Cambridge, MA: MIT Press).

Bolter, Jay D. (2002) 'Formal Analysis and Cultural Critique in Digital Media Theory', *Convergence: The Journal of Research into New Media Technologies* 8 (4), pp. 77–88.

Boorstin, Jon (1995) *Making Movies Work* (Los Angeles: Silman-James Press).

Booth, Wayne (1988) *The Company We Keep: An Ethics of Fiction* (Berkeley: University of California Press).

Brain Sees Violent Video Games as Real Life (2005), http://news.yahoo.com/s/nm/20050622/tc_nm/science_videos_dc_2 [Accessed 2 Nov. 2005].

Bredbeck, Gregory W. (1991) *Sodomy and Interpretation: Marlowe to Milton* (Ithaca: Cornell University Press).

Brown, Jeff of Electronic Arts (2001) 'Microsoft Takes Towers out of Flight Simulator', *San Francisco Chronicle*, 15 September, http://www.chron.com/cs/CDA/ssistory.mpl/special/terror/impact/1050218 [Accessed 10 January 2007].

Burke, Edmund (1998) *A Philosophical Enquiry into the Origin of Our Ideas of the Sublime and Beautiful*, ed. David Womersley (New York: Penguin).

Butler, Judith (1990) *Gender Trouble: Feminism and the Subversion of Identity* (New York: Routledge).

Calvert, Sandra L. (2002) 'Identity Construction on the Internet', in Sandra L. Calvert, Amy B. Jordan, and Rodney R. Cocking (eds) *Children in the Digital Age: Influences of Electronic Media on Development* (Westport, CT: Praeger), pp. 57–70.

Canetti, Elias (1984) *Crowds and Power* (New York: Noonday Press).

Carey, James W. (1988) *Communication as Culture: Essays on Media and Society* (London; Boston: Unwin Hyman).

Castonguay, James (n.d.) 'The Spanish-American War in U.S. Media Culture', Fairfax, VA: Center for History and New Media, George Mason University, http://chnm.gmu.edu/aq/war/index.html [Accessed 23 November 2005].

Castronova, Edward (2001) 'Virtual Worlds: A First-hand Account of Market and Society on the Cyberian Frontier', December 2001, CESifo Working Paper Series No. 618, http://ssrn.com/abstract = 294828 [Accessed 29 May 2006].

Castronova, Edward (2004) 'The Price of Bodies: A Hedonic Pricing Model of Avatar Attributes in a Synthetic World', *Kyklos* 57 (2), pp. 173–96.

Chaplin, Heather and Aaron Ruby (eds) (2005): *Smartbomb. The Quest for Art, Entertainment, and Big Bucks in the Videogame Revolution* (New York: Algonquin Books).

Chen, Kuan-Hsing (1998) 'Introduction: The Decolonization Question', in Kuan Hsing Chen (ed.) *Trajectories: Inter-Asia Cultural Studies* (London; New York: Routledge), pp. 1–53.

China Busy (2004) 'China busy developing homebred online games', *People's Daily Online*, 18 January, http://english1.people.com.cn:80/200401/18/ print20040118_132908.html [Accessed 25 August 2005].

Ching, Leo (2000) 'Globalizing the Regional, Regionalizing the Global: Mass Culture and Asianism in the Age of Late Capital', *Public Culture*, 12 (1), pp. 233–57.

Chou, Yuntsai (2003) 'G-commerce in East Asia: Evidence and Prospects', *Journal of Interactive Advertising* 4 (1), http://jiad.org/vol4/no1/chou/ [Accessed 25 August 2005].

Chua, Beng Huat (2004) 'Conceptualizing an East Asian Popular Culture', *Inter-Asia Cultural Studies* 5 (2), pp. 200–21.

City of Heroes – Game Info: Geography (n.d.), *City of Heroes*, http://www.cityofheroes.com/gameinfo/geography.html [Accessed 13 February 2007].

City of Heroes – Newspaper (n.d.), *City of Heroes*, http://www.cityofheroes.com/paper/newspaper7.html [Accessed 13 February 2007].

Clausewitz, Carl von (1976) *On War,* translated and edited by Michael Howard and Peter Paret (Princeton: Princeton University Press).

Colebrook, Claire (2002) *Understanding Deleuze* (NSW, Australia: Allen and Unwin).

Colombo, Fausto and Ruggero Eugeni (1996) *Il testo visibile* (Rome: La Nuova Italia Scientifica).

Command & Conquer DEN (n.d.), fan website, http://www.cncden.com [Accessed 10 February 2007].

Condon, Brody (2004) *deRez_Fxkill Elvis*, http://tmpspace.com/content/?p=98 [Accessed 10 January 2007].

214   References

ConsoleWire.com Staff (2001) 'Video Games after the Attack', *ConsoleWire.com*, 21 September, http://www.internetnews.com/special/article.php/10716_889361 [Accessed 10 January 2007].

Crary, James W. (1992) *Techniques of the Observer: On Vision and Modernity in the Nineteenth Century* (Cambridge, MA: MIT Press).

Dale, John R. (2002) 'Microsoft Flight Simulator 2002', *iPilot* website, http://www.ipilot.com/feature/review-msfs-02.asp [Accessed 10 January 2007].

Daly, Nicholas (n.d.) 'The Phantom of the Cinema: The Boer War and Early Responses to the Cinematograph', North American Victorian Studies Association (NAVSA), http://www.cla.purdue.edu/academic/engl/navsa/conferences/2003/abstracts/daly.html [Accessed 3 January 2006].

DANBAT_BillyTheKid, Axleonline (n.d.), '"Afghanistan" mission for *Operation Flashpoint*', http://www.axleonline.com/missions/full/Single%20Player/Afghanistan.zip [Accessed 10 October 2003].

Davis, Todd and Kenneth Womack (eds) (2001) *Mapping the Ethical Turn* (Charlottesville: University Press of Virginia).

Deleuze, Gilles and Félix Guattari (1983) *Anti-Oedipus: Capitalism and Schizophrenia*, trans. Robert Hurley, Mark Seem, and Helen R. Lane (Minneapolis: University of Minnesota Press).

Deleuze, Gilles (1987) *A Thousand Plateaus: Capitalism and Schizophrenia*, trans. Brian Massumi (Minneapolis: University of Minnesota Press).

Deleuze, Gilles (1993) *The Fold: Leibniz and the Baroque*, trans. Tom Conley (Minneapolis: University of Minnesota Press).

Deleuze, Gilles (2002) 'Desire and Pleasure', trans. Melissa McMahon, *Interactivist Info Exchange*, 18 November, http://info.interactivist.net/print.pl?sid=02/11/18/1910227 [Acccessed 29 May 2006].

Deleuze, Gilles and Félix Guattari (1994) *What is Philosophy?* (New York: Columbia University Press).

Deleuze, Gilles and Claire Parnet (1987) *Dialogues*, trans. Hugh Tomlinson and Barbara Habberjam (New York: Columbia University Press).

Der Derian, James (ed.) (1998) *The Virilio Reader* (Malden, MA: Blackwell Publishers).

Der Derian, James (2001) *Virtuous War: Mapping the Military-Industrial-Media-Entertainment Network* (Boulder: Westview Press).

De Certeau, Michel (2002) *The Practice of Everyday Life* (Berkeley: University of California Press).

De Peuter, Grieg and Nick Dyer-Witheford (2005) 'Games of Empire: a Transversal Media Inquiry', *Conference Proceedings: Genealogies of Biopolitics*, 18 October, http://www.radicalempiricism.org/biotextes/textes/witheford_peuter.pdf [Accessed 2 May 2006].

Derrida, Jacques (1967) 'Structure, Sign and Play in the Discourses of the Human Sciences', in *Writing and Difference*, (University of Chicago Press), pp. 278–93.

Derrida, Jacques (1989) *Memoires for Paul de Man* (New York: Columbia University Press).

Doležel, Lubomír (1998) *Heterocosmica: Fiction and Possible Worlds* (Baltimore: Johns Hopkins University Press).

Dovey, Jon and Helen W. Kennedy (2006) *Game Cultures: Computer Games as New Media* (Maidenhead: Open University Press).

Dewey, John (1958) *Experience and Nature* (New York: Dover Publications).

Düwell, Marcus, Chistoph Hübenthal, and Micha H. Werner (2002), *Handbuch Ethik* (Stuttgart: Metzler).

Eco, Umberto (1979) *The Role of the Reader* (Bloomington: Indiana University Press).

Elder, Larry (2000) *The Ten Things You Can't Say in America* (New York: St. Martin's Griffin).

Embassy of the Republic of Korea (2004) 'Korean Online Game Developers Flourish in China', *Korea Update* 15 (8), 3 May, http://www.koreaemb.org/archive/2004/5_1/econ/econ4_print.asp [Accessed 19 October 2005].

Engeli, Maia (2005) 'Taxonomy – From a Design Perspective – of Art Game Mods of a Shooter Game or Art Shooter Game Developments', http://maia.enge.li/gamezone/taxonomy.html [Accessed 10 January 2007].

Erdl, Marc F. (2004) *Die Legende von der Politischen Korrektheit. Zur Erfolgsgeschichte eines importierten Mythos* (Bielefeld: transcript).

Eyal, Keren, Miriam J. Metzger, Ryan W. Lingsweiler, Chad Mahood and Mike Z. Yao (2006) 'Aggressive Political Opinions and Exposure to Violent Media', *Mass Communication & Society* 9 (4), pp. 399–428.

Fabulous 999 – Newgrounds (2002) 'The Suicide Bomber Game', *Newgrounds*, 17 April, http://www.newgrounds.com/portal/view.php?id=50323 [Accessed 10 February 2007].

Feldman, Curt (2004) 'China Backs Local Game Developers', *GameSpot*, October 21, http://www.gamespot.com/news/2004/10/21/news_6111054.html [Accessed 19 October 2005].

Ferenczi, Sándor (1952) *First Contributions to Psychoanalysis* (London: Hogarth Press).

First to Fight Website (n.d.), Destineer Studios, http://www.firsttofight.com/html/ [Accessed 10 February 2007].

Fish, Stanley (1982) *Is There a Text in this Class? The Authority of Interpretive Communities* (Cambridge: Harvard University Press).

Foucault, Michel (1999) *Abnormal*, trans. Graham Burchell (New York: Picador).

Fraioli, Alex (2003) 'Koei Invades Singapore', *1UP*, November, http://www.findarticles.com/p/articles/mi_zd1up/is_200311/ai_ziff111594 [Accessed 19 October 2005].

Frankel, Jon (2002) 'Everquest or Evercrack?', *CBS News: The Early Show*, 28 May, http://www.cbsnews.com/stories/2002/05/28/earlyshow/living/caught/main5 10302.shtml [Accessed 2 June 2006].

Frasca, Gonzalo (2001) *Videogames of the Oppressed: Videogames as a Means for Critical Thinking and Debate*, MA Thesis, School of Literature, Communication and Culture, Georgia Institute of Technology.

Freeman, Colin (2005) 'Battles Re-Enacted in Video Arcades', *San Francisco Chronicle*, 16 January: A4.

Freeman, David (2004) *Creating Emotion in Games. The Craft and Art of Emotioneering* (Indiana: New Riders).

Friedman, Ted (1999) '*Civilization* and Its Discontents: Simulation, Subjectivity, and Space', in Greg M. Smith (ed.) *On a Silver Platter: CD-ROMs and the Promises of a New Technology* (New York: New York University Press), pp. 132–50.

Friedman, George (2004) *America's Secret War: Inside the Hidden Worldwide Struggle Between the United States and Its Enemies* (Toronto: Doubleday Canada).

Fullerton, Tracy (2005) 'Documentary Games: Putting the Player in the Path of History', *Playing the Past Conference*, University of Florida, Gainesville, March.

Fulp, Tom, Newgrounds (n.d.) 'Site Background: History of NG', *Newgrounds.com*, http://newgrounds.com/lit/history.html [Accessed 10 May 2006].

Gallagher, David F. (2001) 'Game Makers Scale Back: Players Have Other Ideas', *New York Times*, 20 September; 'Game Changes/Delays Due to WTC Disaster', *elited.net*, 17 September, http://www.elited.net/archives/september.shtml. [Accessed 10 January 2007].

Gallison, Peter (2003) *Einstein's Clocks, Poincaré's Maps: Empires of Time* (New York; London: W.W. Norton & Company).

Galloway, Alexander R. (2004) 'Social Realism in Gaming', *Game Studies* 4 (1). http://www.gamestudies.org/0401/galloway/ [Accessed 10 February 2007].

Galloway, Alexander R. (2006) *Gaming: Essays on Algorithmic Culture* (Minneapolis: University of Minnesota Press).

Gallup Organization (2003) 'Do you think that one man was responsible for the assassination of President Kennedy, or do you think that others were involved in a conspiracy?', Q19 Form A, 10–12 November, http://brain.gallup.com/documents/questionnaire.aspx?STUDY=P0311050 [Accessed 15 December 2003].

Garite, Matt (2003) 'The Ideology of Interactivity (Or Videogames and the Taylorization of Leisure)', *Proceedings of the 2003 Level Up Conference held at the University of Utrecht*, Digital Games Research Association, http://www.digra.org/dl/db/05150.15436 [Accessed 3 January 2007].

Gee, James P. (2003) *What Video Games Have to Teach Us About Learning and Literacy* (New York: Palgrave Macmillan).

Gee, James P. (2005) *Why Video Games Are Good for Your Soul*, Playdate Seminar, Parsons School of Design, New York.

Gill, Tim (1996) 'Introduction Part 2: Playing Games with Computers', in Tim Gill (ed.) *Electronic Children: How Children Are Responding to the Information Revolution* (London: National Children's Bureau).

Giner-Sorolla, Roger (1996) 'Crimes Against Mimesis', USENET post on *rec.arts.interactive-fiction*, April, http://www.geocities.com/aetus_kane/writing/cam.html [Accessed 12 February 2007].

Goldstein, Jeffrey (ed.) (1998) *Why We Watch: The Attractions of Violent Entertainment* (New York: Oxford University Press).

Goldstein, Joshua S. (2001) *War and Gender* (Cambridge: Cambridge University Press).

Greimas, Algirdas J. (1983) *Du Sens II* (Paris: Editions du Seuil).

Greimas, Algirdas J. and Joseph Courtés (1982) *Semiotics and Language* (Bloomington: Indiana University Press).

Grierson, John (1966) *Grierson on Documentary*, ed. Forsyth Hardy (Los Angeles: University of California Press).

Griffiths, Mark (2005) 'The Therapeutic Value of Video Games', in: Joost Raessens and Jeffrey Goldstein (eds) *Handbook of Computer Game Studies* (Cambridge, MA: MIT Press) pp. 161–71.

Grodal, Torben (2000) 'Video Games and the Pleasures of Control', in Dolf Zillmann and Peter Vorderer (eds) *Media Entertainment. The Psychology of Its Appeal* (Mahwah, NJ: Lawrence Erlbaum Associates) pp. 197–214.

Grodin, Debra and Thomas R. Lindlot (eds) (1996) *Constructing the Self in a Mediated World* (London: Sage).

Grossman, Dave (2006) *Killology Research Group*, 1 June, http://www.killology.com [Accessed 3 June 2006].

Grossman, Dave and Gloria DeGaetano (1999) *Stop Teaching Our Kids to Kill: A Call to Action Against TV, Movie and Video Game Violence* (New York: Crown Books).

Grover, Jan Z. (1987) 'AIDS: Keywords', in Douglas Crimp (ed.) *AIDS: Cultural Analysis/Cultural Activism* (Cambridge, MA: MIT Press) pp. 17–30.

Halberstam, Judith (1995) *Skin Shows: Gothic Horror and the Technology of Monsters* (Durham, NC: Duke University Press).

Halberstam, Judith (2000) *Skin Shows: Gothic Horror and the Technology of Monsters* (Durham, NC: Duke University Press).

Hanson, Matt (2004) *The End of Celluloid: Film Futures in the Digital Age* (Hove: RotoVision).

Hartmann, Tilo (2003) 'Gender differences in the use of computer-games as competitive leisure activities', poster presentation at 'Level Up', the 1st Conference on Digital Games, Utrecht, The Netherlands, 4–6 November.

Haynes, Cynthia (2006) 'Armageddon Army: Playing God, God Mode Mods, and the Rhetorical Task of Ludology', *Games and Culture* 1 (1), January.

Heath, Stephen (1981) *Questions of Cinema* (London: Macmillan).

Hedges, Chris (2002) *War is a Force that Gives Us Meaning* (New York: Anchor Books).

Herz, J.C. (1997) *How Videogames Ate Our Quarters, Won Our Hearts, and Rewired Our Minds* (Boston: Little, Brown and Co).

Hillis Miller, Joseph (1987) *The Ethics of Reading* (New York: Columbia University Press).

Hjelmslev, Louis (1961) *Prolegomena to a Theory of Language* (Madison: University of Wisconsin Press).

Holland, Edward W. (1996) 'Schizoanalysis and Baudelaire: Some Illustrations of Decoding at Work', in Paul Patton (ed.) *Deleuze: A Critical Reader* (Cambridge, MA: Blackwell Publishers) pp. 240–56.

Huhtamo, Erkki (1999) 'Game Patch – the Son of Scratch?', *Cracking the Maze*, http://switch.sjsu.edu/CrackingtheMaze/erkki.html [Accessed January 2007].

Huizinga, Johan (1950) *Homo Ludens: A Study of the Play-Element in Culture*, (Boston: The Beacon Press).

Hynes, Samuel (1991) *A War Imagined: The First World War and English Culture* (New York: Atheneum).

Innis, Harold A. (2003) *The Bias of Communication* (University of Toronto Press).

Jahn-Sudmann, Andreas (2006) *Der Widerspenstigen Zähmung? Zur Politik der Repräsentation im gegenwärtigen US-amerikanischen Independent-Film* (Bielefeld: transcript).

Jansz, Jeroen (2005) 'The Emotional Appeal of Violent Video Games for Adolescent Males', *Communication Theory* 15 (3), pp. 219–41.

JasonEthos, Reader Review (2004), *Gamefaqs*, 15 November, http://www.gamefaqs.com/console/ps2/review/R81134.html [Accessed 10 December 2004].

Jenkins, Henry (1992) *Textual Poachers: Television Fans & Participatory Culture* (New York; London: Routledge).

Jenkins, Henry (2002) 'Art Form for the Digital Age', *Technology Review*, 1 November, http://www.technologyreview.com/read_article.aspx?id=12189&ch=infotech [Accessed 10 January 2007].

Jenkins, Henry (2004) 'Game Design as Narrative Architecture', in Noah Wardrip-Fruin and Pat Harrigan (eds) *First Person. New Media as Story, Performance and Game* (Cambridge, MA: MIT Press) pp. 118–130.

Johnson, Steven (1997) *Interface Culture: How New Technology Transforms the Way We Create & Communicate* (San Francisco: Basic Books).

Jordan, John (2003) 'Reality Cheque', *New Scientist* 179 (2405), pp. 44–6.

Juul, Jesper (2001a) 'Editorial', *GameStudies*, http://www.gamestudies.org/ 0501/editorial/ [Accessed 11 January 2007].

Juul, Jesper (2001b): 'A Clash Between Game and Narrative', http://www.jesper-juul.dk/thesis/ [Accessed 11 November 2006].

Kattenbelt, Chiel and Joost Raessens (2003) 'Computer Games and the Complexity of Experience', Proceedings of the 2003 Level Up Conference held at the University of Utrecht, Digital Games Research Association, http://digra.org/dl/ db/05163.48201 [Accessed 3 January 2007].

Kennedy, Harold (2002) 'Computer Games Liven Up Military Recruiting, Training', *National Defense*, November, http://www.nationaldefensemagazine.org/issues /2002/Nov/Computer_Games.htm [Accessed 10 February 2007].

Kern, Stephen (1983) *The Culture of Time and Space 1880–1918* (Cambridge, MA: Harvard University Press).

Kilgannon, Corey (2002) 'Wanted, Virtually Dead', *New York Times*, 10 January, http://tech2.nytimes.com/mem/technology/techreview.html?res=9E07E6D71 339F933A25752C0A9649C8B63 [Accessed 10 February 2007].

Kilian, Monika (1998) *Modern and Postmodern Strategies: Gaming and the Question of Morality* (New York: Peter Lang Publishing).

Kinder, Marsha (2002) 'Narrative Equivocations between Movies and Games', in Dan Harries (ed.) *The New Media Book* (London: BFI) pp. 119–32.

Kim, Jung Ryul (2004a) *Online Worlds Roundtable #11, Part 1*, 16 August, http://rpg-vault.ign.com/articles/539/539073p2.html [Accessed 25 August 2005].

Kim, Jung Ryul (2004b) *Online Worlds Roundtable #11, Part 3*, 3 September, http://rpgvault.ign.com/articles/544/544318p3.html [Accessed 25 August 2005].

Kirsh, Steven J. (2003) 'The Effects of Violent Video Games on Adolescents: The Overlooked Influence of Development', *Aggression and Violent Behavior* 8, pp. 377–89.

Klimmt, Christoph (2003) 'Dimensions and Determinants of the Enjoyment of Playing Digital Games: A Three-level Model', in Marinka Copier and Joost Raessens (eds) *Level Up: Digital Games Research Conference*, Utrecht: Faculty of Arts, Utrecht University, pp. 246–57.

Klimmt, Christoph and Sabine Trepte (2003) 'Theoretisch-methodische Desiderata der medienpsychologischen Forschung über die aggressionsfördernde Wirkung gewalthaltiger Computer- und Videospiele', *Zeitschrift für Medienpsychologie* 15 (4), pp. 114–21.

Klimmt, Christoph and Peter Vorderer (2003) 'Media Psychology "Is Not Yet There": Introducing Theories on Media Entertainment to the Presence Debate', *Presence: Teleoperators and Virtual Environments* 12 (4), pp. 346–59.

Klimmt, Christoph, Hannah Schmid, Andreas Nosper, Tilo Hartmann and Peter Vorderer (2006) 'How Players Manage Moral Concerns To Make Video Game Violence Enjoyable', *Communications – the European Journal of Communication Research* 31 (3), pp. 309–28.

Kofman, Amy Ziering and Kerby Dick (2005) *Derrida* [DVD] (London: ICA Projects).

Korea Game Development and Promotion Institute (KGDI) (2004) *2004 The Rise of Korea Games* (Seoul: KGDI).

Krzywinska, Tanya (2006) 'The Pleasures and Dangers of the Game: Up Close and Personal', *Games and Culture* 1 (1), January, pp. 119–22.

Kuhrcke, Tim, Christoph Klimmt and Peter Vorderer (2005, submitted for presentation) *Why Is Virtual Fighting Fun? Motivational Predictors of Exposure to Violent Video Games*, paper submitted to the Annual Conference of the International Communication Association (ICA), June 19–23, 2006, Dresden.

Kushner, David (2003) *Masters of Doom: How Two Guys Created an Empire and Transformed Pop Culture* (New York: Random House).

LaFarge, Antoinette (2000) *Winside Out – an Introduction to the Convergence of Computers, Games, and Art*, http://beallcenter.uci.edu/shift/essays/lafarge.html [Accessed January 2007].

Landow, George (1992) *Hypertext* (Baltimore: Johns Hopkins University Press).

Landow, George P. (ed.) (1994) *Hyper/Text/Theory* (Baltimore: Johns Hopkins University Press).

Landow, George P. (1997) *Hypertext 2.0: The Convergence of Contemporary Critical Theory and Technology* (Baltimore: Johns Hopkins University Press).

Lee, Keng-Fang (2003) 'Far Away, So Close: Cultural Translation in Ang Lee's *Crouching Tiger, Hidden Dragon'*, *Inter-Asia Cultural Studies* 4 (2), pp. 281–94.

Leonard, David J. (2006) 'Not a Hater, Just Keeping It Real. The Importance of Race- and Gender-Based Game Studies', *Games and Culture* 1 (1), January, pp. 83–8.

Levander, Michelle (2001) 'Where Does Fantasy End?' *Time*, 157 (22), June 4, http://www.time.com/time/interactive/entertainment/gangs_np.html [Accessed 25 August 2005].

Lévi-Strauss, Claude (1983) *The Raw and the Cooked: Mythologiques* 1 (Chicago: University of Chicago Press).

Levy, Pierre (2001) *Cyberculture* (Minneapolis: University of Minnesota Press).

Levy, Steven (1984) *Hackers: Heroes of the Computer Revolution* (Garden City: Anchor Press).

Lin, Frances (2002) 'The Multiplicity of Taiwan's Online Games Market', http://www.tdctrade.com/imn/02071102/info31.htm [Accessed 19 October 2005].

Liu, Alexandra (2001) 'Flat Screens and Flying Fists: Martial Arts Gaming in Taiwan', *Sinorama Magazine*, http://www.sinorama.com.tw/en/print_issue. php3?id=2001109010032e.txt&mag=past [Accessed 25 August 2005].

Lowood, Henry and Timothy Lenoir (2005) 'Theaters of War: The Military-Entertainment Complex', in Helmar Schramm, Ludger Schwarte and Jan Lazardzig (eds) *Collection – Laboratory – Theater: Scenes of Knowledge in the 17th Century* (Berlin; New York: Walter de Gruyter), pp. 427–56.

MacCabe, Colin (1981) 'Realism: Notes on Some Brechtian Theses', in Tony Bennett (ed.) *Popular Television and Film* (London: BFI) pp. 216–36.

Macedonia, Michael (2002) 'Computer Games and the Military: Two Views – A View from the Military', *Defense Horizons* 11, pp. 6–8.

Malins, Peta (2004) 'Machinic Assemblages: Deleuze, Guattari and an Ethico-Aesthetics of Drug Use', *Janus Head* 7 (1), pp. 84–104.

Manovich, Lev (2001) *The Language of New Media* (Cambridge, MA: MIT Press).

Manske, Ariane (2002) *Political Correctness und Normalität. Die amerikanische PC-Kontroverse im kulturgeschichtlichen Kontext* (Heidelberg: Synchron).

Marks, Robert B. (2003) *EverQuest Companion: The Inside Lore of a Gameworld* (Emeryville, CA: McGraw-Hill/ Osborne).

Marwil, Jonathan (2000) 'Photography at War', *History Today* 50 (6), pp. 30–7.

McArthur, Tom (1986) *Worlds of Reference: Lexicography, Learning and Language from the Clay Tablet to the Computer* (Cambridge: Cambridge University Press).

McClellan, Jim (2004) 'The Role of Play', *The Guardian*, 13 May, http://technology.guardian.co.uk/online/story/0,3605,1214955,00.html [Accessed 12 February 2007].

McLuhan, Marshall (1962) *The Gutenberg Galaxy; The Making of Typographic Man* (Toronto: University of Toronto Press).

McLuhan, Marshall (1994) *Understanding Media: The Extensions of Man* (Cambridge, MA: MIT Press).

McLuhan, Marshall (1997) *Understanding Media: The Extensions of Man*, 5th printing (Cambridge, MA: MIT Press).

Microsoft (2003) 'Xbox Expands Definition of Digital Entertainment Lifestyle With New and Ambitious Titles Featured at E3', press release 12 May, http://www.microsoft.com/presspass/press/2003/may03/05-12e3gamespr.mspx [Accessed 12 February 2007].

Mighty Quasar (2001) 're: new plane, headed towards washington', *Allspark.net forum*, 11 September, http://p084.ezboard.com/ftheallsparkfrm47.showMessage?topicID=9.topic [Accessed 10 January 2007].

Miller, Stanley A. II and Joe Winter (2002) 'Death of a Game Addict', *Milwaukee Journal-Sentinel*, 31 March, http://www2.jsonline.com/news/state/mar02/31536.asp [Accessed 13 May 2006].

Mirapaul, Matthew (2003) 'Online Games Grab Grim Reality', *New York Times*, 17 September, Arts and Culture section.

Morse, Margaret (1998) *Virtualities: Television, Media Art, and Cyberculture* (Bloomington: Indiana University Press).

Mosher, Donald L. and Mark Sirkin (1984) 'Measuring a Macho Personality Constellation', *Journal of Research in Personality* 18, pp. 150–63.

Mr. Domino (2000) 'Rainbow Six: Rogue Spear', *Planet Dreamcast*, 7 September, http://www.planetdreamcast.com/games/reviews/rainbowsixroguespear/ [Accessed 12 February 2007].

Müller-Lietzkow, Jörg, Ricarda B. Bouncken, and Wolfgang Seufert (2006) *Gegenwart und Zukunft der Computer- und Spieleindustrie* (Dornach bei München).

Mulvey, Laura (1975) 'Visual Pleasure and Narrative Cinema', *Screen* 16 (3), pp. 6–18.

Mundt, Michaela (1994) *Transformationsanalyse. Methodologische Probleme der Literaturverfilmung* (Tübingen: Niemeyer Verlag).

Murray, Janet (1997) *Hamlet on the Holodeck: The Future of Narrative in Cyberspace* (Cambridge, MA: MIT Press).

Musser, Charles (1990) *The Emergence of Cinema: The American Screen to 1907* (New York: Scribner).

Neitzel, Britta and Rolf F. Nohr (eds) (2006) *Das Spiel mit dem Medium. Partizipation – Immersion – Interaktion. Zur Teilhabe an den Medien von Kunst bis Computerspiel* (Marburg: Schüren).

Newman, James (2004) *Videogames* (London; New York: Routledge).

Nichols, Bill (1991) *Representing Reality: Issues and Concepts in Documentary* (Bloomington: Indiana University Press).

Nichols, Bill (2001) *Introduction to Documentary* (Bloomington: Indiana University Press).

Nietzsche, Friedrich (1990) *Beyond Good and Evil: Prelude to a Philosophy of the Future*, trans. R. J. Hollingdale (London: Penguin).

Nirrian and Keeter (2001), 're: Candlelight Vigil on Luclin', Everlore Message Board, 12 September, http://www.everlore.com [Accessed 10 May 2006].

Nutt, Diane and Diane Railton (2001) '*The Sims*: Real Life as Genre', *Information, Communication and Society* 6 (4), pp. 577–92.

On, Josh (n.d.) 'Net Art and Games as Protest Media', interview by Robert Praxmarer, http://www.lowfi.org.uk/rethinkingwargames/docs/interview_rob prax.htm#interview [Accessed 12 February 2007].

Ong, Walter J. (1982) *Orality and Literacy: The Technologization of the Word* (New York: Methuen).

PainStation Website (n.d.), http://www.painstation.de [Accessed 10 February 2007].

Pavel, Thomas (1989) *Fictional Worlds* (Cambridge, MA: Harvard University Press).

Pearce, Celia (2004) 'Towards a Game Theory of Game' in Noah Wardrup-Fruin and Pat Harrigan (eds) *First Person: New Media as Story, Performance and Game* (Cambridge, MA: MIT Press) pp. 143–53.

Perlin, Ken (2004) 'Can There Be a Form between a Game and a Story?', in Noah Wardrip-Fruin and Pat Harrigan (eds) *First Person: New Media as Story, Performance and Game* (Cambridge, MA: MIT Press) pp. 12–18.

Perron, Bernard and Mark J.P. Wolf (eds) (2003) *The Video Game Theory Reader* (New York; London: Routledge).

Pilla, Matt (2001) 'Microsoft Takes Towers out of Flight Simulator', *San Francisco Chronicle*, 15 September, http://www.chron.com/cs/CDA/ssistory.mpl/special/terror/impact/1050218 [Accessed 10 January 2007].

Plotnitsky, Arkady (1994) *Complementarity* (Durham, NC: Duke University Press).

Poole, Steven (2000) *Trigger Happy. Videogames and the Entertainment Revolution* (New York: Arcade Publishing).

Potter, W. James and Tami K. Tomasello (2003) 'Building Upon the Experimental Design in Media Violence Research: The Importance of Including Receiver Interpretations', *Journal of Communication* 53, pp. 315–29.

Prensky, Marc (2001) *Digital Game-Based Learning* (New York; London: McGraw-Hill).

Protevi, John (2001) 'The Organism as the Judgment of God: Aristotle, Kant, and Deleuze on Nature (That Is, on Biology, Theology and Politics)', in Mary Bryden (ed.) *Deleuze and Religion* (London; New York: Routledge) pp. 30–41.

Provenzo, Eugene (1991) *Video Kids: Making Sense of Nintendo* (Cambridge, MA: Harvard University Press).

Quaranta, Domenico (2006) 'Julian Oliver aka Delire, q3aPaint', in Matteo Bittani and Domenico Quaranta (eds) *Gamescenes – Art in the Age of Videogames* (Milan: Johan & Levi Editore).

Raessens, Joost and Jeffrey Goldstein (eds) (2005) *Handbook of Computer Game Studies* (Cambridge, MA: MIT Press).

Raessens, Joost (2006) 'Playful Identities or the Ludification of Culture', *Games and Culture* (1) 1, January, pp. 52–7.

Ragaini, Toby (2004) *Online Worlds Roundtable #11, Part 3*, 3 September, http://rpgvault.ign.com/articles/544/544318p3.html [Accessed 25 August 2005].

Red Storm Website (n.d.), http://www.redstorm.com [Accessed 10 February 2007].

Red Storm Website (2001), 'Ubisoft Postpones Release of Tom Clancy's *Rogue Spear: Black Thorn*', news release, 17 September, http://www.redstorm.com/blackthorn/fullpost.php?id=76 [Accessed 10 February 2007].

Rein-Hagen, Mark (1992) *Vampire – The Masquerade* (Stone Mountain: White Wolf).

Reynolds, Ren (2002) 'Playing a "Good" Game: A Philosophical Approach to Understanding the Morality of Games', *International Game Developers Association*, http://www.igda.org/articles/rreynolds_ethics.php [Accessed 12 February 2007].

Roberts, Donald F., Ulla G. Foehr and Victoria J. Rideout (2005) *Generation M: Media in the Lives of 8–18-year-olds* (Menlo Park: Kaiser Family Foundation).

Rombes, Nicholas (2004) 'Blue Velvet Underground: David Lynch's Post-Punk Poetics', in Erica Sheen and Annette Davison (eds) *The Cinema of David Lynch: American Dreams, Nightmare Visions* (London: Wallflower) pp. 61–76.

Rossiter, Ned (2003) 'Processual Media Theory', *Symploke* 11 (1–2), pp. 104–31.

Saltzman, Mark (2001) 'Army Enlists Simulation to Help Tackle Terrorists', *USA Today*, 2 October.

Santos, Marc C., and Sarah E. White (2005) 'Playing with Ourselves: a Psychoanalytic Investigation of *Resident Evil* and *Silent Hill*', in Nate Garrelts (ed.) *Digital Gameplay: Essays on the Nexus of Game and Gamer* (Jefferson: McFarland Press) pp. 69–79.

Scarry, Elaine (1985) *The Body in Pain: The Making and Unmaking of the World* (New York: Oxford University Press).

Schachtman, Noah (2001) 'New Army Soldiers: Game Gamers', *Wired News*, 29 October, http://www.wired.com/news/conflict/0,2100,47931,00.html [Accessed 10 February 2007].

Schiesel, Seth (2005) 'On Maneuvers with the Army's Game Squad', *New York Times*, February 17, The Late Edition, Section G, Column 3, p. 1.

Schivelbusch, Wolfgang (1986) *The Railway Journey: The Industrialization of Time and Space in the 19th Century* (Berkeley: University of California Press).

Schleiner, Anne-Marie (1999) *Cracking the Maze – Game Plug-Ins and Patches as Hacker Art*, online exhibition, http://switch.sjsu.edu/CrackingtheMaze/ [Accessed January 2007].

Schleiner, Anne-Marie (2002) About: [Velvet-Strike], website, http://www.opensorcery.net/velvet-strike/ [Accessed 10 February 2007].

Schneider, Edd F., Annie Lang, Mija Shin and Samuel D. Bradley (2004) 'Death with a Story: How Story Impacts Emotional, Motivational, and Physiological Responses to First-person Shooter Video Games', *Human Communication Research* 30 (3), pp. 361–75.

Schwarz, Daniel (2001) 'A Humanistic Ethics of Reading', in Todd Davis and Kenneth Womack (eds) *Mapping the Ethical Turn* (Charlottesville: University Press of Virginia).

Semetsky, Inna (2004) 'The Problematics of Human Subjectivity: Gilles Deleuze and the Deweyan Legacy', *Studies in Philosophy & Education* 22 (3), pp. 211–25.

Sherry, John L. (2001) 'The Effects of Violent Video Games on Aggression. A Meta-analysis', *Human Communication Research* 27 (3), pp. 409–31.

Shlain, Leonard (1998) *The Alphabet Versus the Goddess* (New York: Viking).

Shotter, John, and Kenneth J. Gergen, (eds) (1989) *Texts of Identity* (London: Sage).

Sicart, Miguel (2005) 'The Ethics of Computer Game Design', paper presented at the International DiGRA Conference 2005, *Changing Views: Worlds in Play*, http://www.digra.org/dl/db/06276.55524.pdf [Accessed 12 February 2007].

Sieberg, Daniel (2001) 'War Games: Military Training Goes High-Tech', *CNN.com/ SCI-TECH*, http://archives.cnn.com/2001/TECH/ptech/11/22/war.games/ [Accessed 2 November 2006].

Simmel, Georg (1990) *The Philosophy of Money* (London; New York: Routledge).

Šišter, Vit (2005) 'Videogames and Politics', Entermultimediale 2, Prague, Czech Republic, 9–12 May.

Slater, Michael D. (2003) 'Alienation, Aggression, and Sensation Seeking as Predictors of Adolescent Use of Violent Film, Computer, and Website Content', *Journal of Communication* 53 (1), pp.105–21.

Slater, Michael D., Kimberly L. Henry, Randall C. Swaim and Lori L. Anderson (2003) 'Violent Media Content and Aggressiveness in Adolescents: A Downward Spiral Model', *Communication Research* 30 (6), pp. 713–36.

Slater, Michael D., Kimberly L. Henry, Randall C. Swaim and Joe M. Cardador (2004) 'Vulnerable Teens, Vulnerable Times: How Sensation Seeking, Alienation, and Victimization Moderate the Violent Media Content-aggressiveness Relation', *Communication Research* 31 (6), pp. 642–68.

Smith, Stacy L., Ken Lachlan, and Ron Tamborini (2003) 'Popular Video Games: Quantifying the Presentation of Violence and Its Context', *Journal of Broadcasting and Electronic Media* 47 (1), pp. 58–76.

Smith, Stacy L., Ken Lachlan, Katherine M. Pieper, Aaron R. Boyson, Barbara J. Wilson, Ron Tamborini and Rene Weber (2004) 'Brandishing Guns in American Media: Two Studies Examining How Often and in What Context Firearms Appear on Television and in Popular Video Games', *Journal of Broadcasting and Electronic Media* 48 (4), pp. 584–606.

Smither, Roger (1993) 'A Wonderful Idea of the Fighting: The Question of Fakes in *The Battle of the Somme*', *Historical Journal of Film, Radio and Television* 13 (2), pp. 148–68.

Smuts, Aaron (2005) 'Are Video Games Art?', *Contemporary Aesthetics*, http://www.contempaesthetics.org/newvolume/pages/article.php?articleID=299 [Accessed 5 January 2007].

Sparks, Glenn G. and Cheri W. Sparks (2000) 'Violence, Mayhem, and Horror', in Dolf Zillmann and Peter Vorderer (eds) *Media Entertainment: The Psychology of Its Appeal* (Mahwah, NJ: Lawrence Erlbaum Associates) pp. 73–92.

Stivale, Charles (1997) 'L'Abécédaire de Gilles Deleuze, avec Claire Parnet – An Overview', http://www.langlab.wayne.edu/CStivale/D-G/ABC2.html [Accessed 15 February 2006].

Stockwell, Steven and Adam Muir (2003) 'The Military-Entertainment Complex: A New Facet of Information Warfare', *Fibreculture* 1, http://journal.fibreculture.org/issue1/issue1_stockwellmuir.html [Accessed 24 May 2006].

Stone, Allucquère R. (2000) 'Will the Real Body Please Stand Up?: Boundary Stories about Virtual Cultures', in David Bell and Barbara Kennedy (eds) *The Cybercultures Reader* (New York: Routledge) pp. 504–28.

Suits, Bernard (1978) *The Grasshopper: Games, Life and Utopia* (Toronto: University of Toronto Press).

Sutton-Smith, Brian (1997) *The Ambiguity of Play* (Cambridge, MA: Harvard University Press).

Tamborini, Ron and Paul Skalski (2006) 'The Role of Presence in the Experience of Electronic Games', in Peter Vorderer and Jennings Bryant (eds) *Playing Video Games: Motives, Responses, Consequences* (Mahwah, NJ: Lawrence Erlbaum Associates) pp. 225–40.

Tan, Ed S. (1996) *Emotion and the Structure of Narrative Film. Film as an Emotion Machine* (Mahwah, NJ: Lawrence Erlbaum Associates).

Tan, Terence (2004) *Online Worlds Roundtable #11, Part 1*, 26 August, http://rpgvault.ign.com/articles/539/539073p2.html [Accessed 25 August 2005].

Thacker, Eugene (2004) *Biomedia* (Minneapolis: University of Minnesota Press).

The Great Chicago Fire and The Web of Memory (1996), virtual exhibition, Chicago Historical Society and the Trustees of Northwestern University, http://www.chicagohs.org/fire/media/pic0470.html [Accessed 10 February 2007].

Thompson, Adam J. (2005) 'Morality Play – Creating Ethics in Video Games', http://www.adamjthompson.com/thought/CreatingEthics.html [Accessed 10 December 2005].

Thompson, Clive (n.d.) 'Game Theories', *The Walrus*, http://www.llts.org/articles/walrus/walrus.html [Accessed 13 March 2006].

Thompson, Clive (2002) 'Dot-Columnist: Online Video Games are the Newest Form of Social Comment', *Slate*, 29 August, http://slate.msn.com/?id=2070197 [Accessed 10 February 2007].

Thompson, Clive (2004a) 'Playing Politics: How Dean's Web Game is Like Pac-Man', *Slate*, 16 January, http://slate.msn.com/id/2094039 [Accessed 10 February 2007].

Thompson, Clive (2004b) 'The Making of an X-Box Warrior', *The New York Times Magazine*, 22 August, pp. 35–6.

Trotter, William (2003) 'The Power of Simulation: Transforming Our World', *PC Gamer*, February, pp. 24–8.

Tuchscherer, Pamela (1988) *The New High Tech Threat to Children* (Bend: Pinnaroo Publishing).

Turkle, Sherry (2003) 'Video Games and Computing Holding Power', in Noah Wardrip-Fruin and Nick Monfort (eds) *The New Media Reader* (Cambridge, MA: MIT Press) pp. 409–514.

Ubi Soft press release (2001), 'Ubi Soft Licenses Rogue Spear Game Engine to Train U.S. Soldiers', 29 August.

Under Ash Website (2004), http://www.underash.net/en_download.htm [Accessed 10 February 2007].

Upton, Brian (1999) 'Postmortem: Red Storm's *Rainbow Six*', *Game Developer Magazine*, May.

Upton, Brian (2004) 'Re: Which Type of Writer Makes the Best Game Writer?', IGDA Writing Forum, 22 June, http://www.igda.org/Forums/showthread.php?postid=69183 [Accessed 10 February 2007].

Virilio, Paul (1989) *War and Cinema: The Logistics of Perception* (London; New York: Verso).

Vorderer, Peter and Silvia Knobloch (2000) 'Conflict and Suspense in Drama', in Dolf Zillmann and Peter Vorderer (eds) *Media Entertainment. The Psychology of Its Appeal* (Mahwah, NJ: Lawrence Erlbaum Associates) pp. 59–73.

Wakin, Daniel J. (2003) 'Video Game Mounts Simulated Attacks Against Israeli Targets', *New York Times*, 18 May, http://www.genocidewatch.org/LebanonHezbollahvideogameMay18.htm [Accessed 10 February 2007].

Walker, Jesse (2003) 'Video Games, Art, and Moral Panic', *Reason*, July, http://www.reason.com/news/show/28834.html [Accessed 12 February 2007].

Wardrip-Fruin, Noah and Nick Monfort (eds) (2003) *The New Media Reader* (Cambridge, MA: MIT Press).

Wardrip-Fruin, Noah and Pat Harrigan (eds) (2004) *First Person: New Media as Story, Performance and Game* (Cambridge, MA: MIT Press).

Wargamer Website (2002) 'G-2: Gaming News and Intel', January, http://www.wargamer.com/news/arc0-2002.htm [Accessed 10 February 2007].

Weber, René, Ute Ritterfeld, and Kostygina, Anna (2006) 'Aggression and Violence as Effects of Playing Violent Video Games?', in Peter Vorderer and Jennings Bryant (eds) *Playing Video Games: Motives, Responses, Consequences* (Mahwah, NJ: Lawrence Erlbaum Associates) pp. 347–62.

Werde, Bill (2004) 'The War at Home', *Wired*, March, pp. 104–5.

Whissel, Katherine (2002) 'Placing the Spectator on the Scene of History: The Battle Re-enactment at the Turn of the Century, from Buffalo Bill's Wild West to the Early Cinema', *Historical Journal of Film, Radio and Television* 22 (3), pp. 225–43.

Wild, Kate et al. (n.d.) *Escape from Woomera*, http://www.escapefromwoomera.com [Accessed January 2007].

Williams, George T. (1999) 'Reluctance to Use Deadly Force', *FBI Law Enforcement Bulletin* 68 (10), pp. 1–5.

Willis, Sharon (1997) *Race and Gender in Contemporary Hollywood Cinema* (Durham, NC; London: Duke University Press).

Winckelmann, Johann J. (1996) 'The History of Ancient Art', in Anne Mellor and Richard Matlak (eds) *British Literature 1780–1830* (Boston: Heinle and Heinle) pp. 129–30.

Winter, Joe (2002) 'Hudson Mother Seeks Answers after Son Addicted to Computer Game Shoots Himself', *RiverTowns.net (Hudson Star Observer)*, 20 February, http://www.rivertowns.net/daily/hso/c020220/#s020220a [Accessed 3 May 2006].

Wolf, Mark J.P. and Bernard Perron (eds) (2003) *The Video Game Theory Reader* (London; New York: Routledge).

Wright, Evan (2004) *Generation Kill: Devil Dogs, Iceman, Captain America, and the Face of American War* (New York: Berkeley Caliber).

Xinhua News Agency. (2003) 'Chinese-made Online Games Take Off', http://www.china.org.cn/english/scitech/82661.htm [Accessed 25 August 2005].

Yu, Eileen (2005) 'Koei Unveils US1.8m Game Plan for Singapore', *CNETAsia*, 15 February, http://asia.cnet.com/news/personaltech/0,39037091,39217780,00.htm [Accessed 19 October 2005].

Zilbergeld, George (2004): *A Reader for the Poltically Incorrect* (Westport, Conn.; London: Praeger).

Zillmann, Dolf (1988) 'Mood Management Through Communication Choices', *American Behavioral Scientist* 31, pp. 327–40.

Zimmerman, Eric and Katie Salen (2004) *Rules of Play* (Cambridge, MA: MIT Press).

Žižek, Slavoj (2002) *The Art of the Ridiculous Sublime: On David Lynch's* Lost Highway (Seattle: Walter Chapin Simpson Center for the Humanities).

Index

9/11, v, vi, xvii, 72, 77–86, 140, 141, 143, 146–9, 211

Aarseth, Espen, 3, 4, 30, 31, 211
aggression, xiv, 54, 108–10, 113, 117, 118, 176, 211, 215, 218, 222, 223, 225
Althusser, Louis, xix, 175, 211
anger, xiii, 14, 22, 28, 41, 65, 71, 83, 86, 95, 112, 125–7, 155, 161, 171, 219
art, ix, xi, xvi, xviii, 8, 13, 20, 35, 36, 39, 40, 63, 78, 87, 88, 96, 151, 163–5, 171, 211, 213, 215, 217, 219–25
artificial, 35, 41, 69, 74, 92, 164
artist, XV, 6, 16, 32, 36, 39–42, 63, 84, 85, 89, 162–9, 171, 172
artwork, 40, 162, 164–6
avatar, 6, 7, 10, 61–3, 65, 105, 119–28, 131, 133, 135, 138–44, 153–55, 166, 187, 190, 197, 203, 206, 213

balancing, 32, 42, 48, 50, 123, 148, 158
Benjamin, Walter, 8, 75, 212
Bolter, Jay D., 145, 176, 178, 212
borders, vi, 63, 140–3, 145, 147, 149

cheats, 11, 36, 125, 156, 157, 160, 201, 205
cinema, 6, 8, 11, 13, 14, 16, 18, 20, 25, 26, 34, 41, 42, 49, 75, 79, 80, 86, 88–91, 93, 96, 105, 124, 132, 153–6, 160, 175–8, 211, 212, 214, 217, 218, 220, 222, 224, 225
Columbine, xiii, 10, 56, 57, 64, 71
consistency, v, 9, 47, 49, 51–5, 63, 123, 151
console, ix, xiii, 7, 40, 54, 78, 84, 140, 167, 169, 170, 188, 190, 214, 217
crime, 154, 163, 198, 211, 216
culture, xi, xiv, xviii, 3, 8, 21, 39, 64, 73, 96, 103, 134, 136, 138, 140, 141, 147, 151, 153–6, 158–60, 163, 178–80, 187, 188, 190–93, 199, 212–21, 223

Deleuze, Gilles, v, xvii, 56–63, 65, 213, 214, 217, 219, 221–3
Derrida, Jacques, 178–82, 184, 212, 214, 218
documentary games, v, xi, 21, 72, 89–91, 215, 216, 220

economics, xiii, xv, 33, 34, 59, 84, 93, 131–4, 137, 189
emergence, 5, 8, 11, 24, 27, 144, 177, 187, 195, 220
emotion, v, x, xii xvi, 14, 22–5, 27–31, 34, 37, 38, 42–4, 49, 72, 76, 78, 79, 82, 83, 99, 108, 117–19, 197, 215, 217, 222, 224
enemy, 90, 116, 138, 139, 141, 157, 170, 207
ethics, vi, xii, xvii, 18, 97, 99–107, 110, 112, 114, 116, 118–24, 126–8, 212, 217, 222, 224
ethnicity, 138, 151–3, 155, 157, 160, 208

fear, xviii, 15, 28, 38, 41, 43, 71, 131, 135–7, 144, 146–8
fiction, v, xii, 17, 22–7, 29–31, 41, 47, 53, 54, 57, 61, 87, 91, 99–107, 144, 148, 155, 212, 214, 216, 221
flow, 39, 40, 58, 59, 63, 102, 177, 195, 201, 222
Foucault, Michel, 138, 144, 215
Frasca, Gonzalo, 3, 10, 85, 215
fun, viii, x, 4, 6, 14, 24–6, 28, 29, 33, 34, 37, 39, 42, 47, 50, 51, 55, 57, 62, 63, 69–71, 73, 75, 77, 81, 82, 84, 99–101, 106, 111, 115–8, 137, 138, 141, 145, 148, 151, 156, 164, 179, 181, 187, 195, 197, 198, 203, 208, 219

game advertising, 61, 71, 81, 83, 93,
    135, 163, 213
game controls, 48, 106, 176
game design, 1, 4–6, 8, 10, 14, 16, 18,
    20, 24, 26, 28, 30, 32, 34, 36, 38,
    40, 42, 44, 211, 217, 222
game experience, 3, 17, 27, 42, 43,
    115, 168
game graphics, xviii, 5, 7, 11, 13, 16,
    35, 39–42, 48, 55, 57, 84, 88, 89,
    91, 94, 124, 140, 156, 157, 159,
    162, 168, 169, 171, 182, 191,
    199–201, 220
game marketing, vi, xi, xviii, 19, 34,
    36, 49, 55, 160, 186, 187, 192,
    196
game mechanics, 3, 8, 9, 11, 50, 123,
    168
game play, vi, vii, xviii, 3, 23, 27, 34,
    71, 77, 110, 114, 117, 118, 173,
    176, 178, 180, 182, 184, 188, 190,
    192, 194, 196, 198, 200, 202, 204,
    206, 222
game sales, xiii, 83, 151, 163, 189
game studies, viii, xi, xv, xvi, xviii,
    150, 216, 219
game theory, v, 32, 33, 35, 37, 39, 41,
    43, 175, 178, 221, 225
game world, xviii, 10, 27, 30, 47, 50,
    51, 57, 58, 64, 79, 99, 101, 107,
    113, 115, 119, 132, 133, 137,
    153–5, 163, 164, 166, 169, 170,
    191
gender, 11, 37, 56, 61, 105, 110, 111,
    125, 143–5, 153, 158, 160, 166,
    172, 197, 201, 202, 206, 213, 216,
    217, 219, 225
geography, 50, 52, 146, 154, 199,
    213
goal, xv, 24, 26–8, 30, 32, 40, 42, 43,
    64, 86, 87, 92, 102, 106, 141, 163,
    165
government, 90, 116, 138, 188, 189

Halberstam, Judith, 127, 144, 146,
    217
hate, 58, 60, 61, 65, 74, 146–8, 157,
    160, 204, 219
Huizinga, Johan, 3, 100, 217

identity, vi, xix, 61, 62, 64, 100, 101,
    105–7, 110, 111, 140, 141, 143,
    146, 147, 151, 155, 162, 175,
    177–80, 183, 184, 186, 187, 192,
    193, 197, 202, 213, 222
immersion, ix, xvii, xix, 41, 47, 50,
    54, 78, 82, 85, 95, 141, 162, 175,
    176, 220
industry, xiii, xvii, 32–4, 40, 78–80,
    83, 84, 92, 93, 95, 136, 154, 161,
    187, 189, 193, 212
innovation, v, ix, xiii, xv, 26, 32–7, 39,
    41, 43, 84, 118
internet, xiv, 72, 80–2, 102, 128, 142,
    157, 169, 189, 190, 213, 214
involvement, 18, 26, 33, 43, 57, 60,
    62–4, 69, 78, 104, 110, 114, 123,
    126, 137, 149, 152, 158, 159, 167,
    170, 183, 191, 194, 198, 199, 201,
    204, 216

Jenkins, Henry, v, x, xv, 9, 24, 32, 155,
    197, 217
Juul, Jesper, xv, 7, 22, 218

Landow, George P., 103, 180, 181,
    219
love, 22, 34, 48, 65, 87, 90, 119, 126,
    136, 141, 145, 150, 159, 164, 197,
    200–2
ludic, 38, 50, 51, 54, 183
ludology, 217

magic circle, 100
McLuhan, Marshall, 5, 73, 77, 163,
    164, 220
media, 3–5, 7–17, 19–24, 26, 30, 32,
    35, 36, 38, 56–58, 69–71, 73–5,
    77–80, 83, 84, 86, 89, 92, 93, 95,
    96, 99, 100, 102, 105, 107,
    109–11, 115, 117, 118, 133, 135,
    145, 146, 150, 153, 154, 156, 157,
    160, 163–5, 175–80, 189, 191,
    202, 211–25
military, vi, x, xvii, 72, 77, 81–7, 89,
    91–5, 108, 116, 133, 138, 142,
    147, 169, 214, 218, 219, 223
MMORPG, vi, xiii, xvii, 19, 58, 140–2,
    147, 186–8, 190–4, 196

morality, vi, xvi, 18, 57, 61, 79, 82, 97,
    99–102, 104, 106, 108–18, 120–2,
    124–8, 132, 133, 138, 162, 163,
    206, 211, 218, 222, 224
multiplayer, ix, xiii, 23, 55, 58, 79,
    140, 169, 187, 188
murder, 43, 56, 111, 121, 123, 124,
    127, 138, 153, 156, 203
Murray, Janet, 3, 6, 29, 75, 142, 148,
    149, 220
music, x, 5, 32, 34, 38, 40, 41, 171

narration, xvii, xix, 3, 8–11, 15, 22,
    24–8, 30, 31, 47, 49–51, 53–5, 57,
    58, 62, 69, 73–5, 83, 87, 89–91,
    99–107, 114–6, 120, 123, 133,
    141–4, 146–9, 159, 177, 179, 186,
    187, 191, 193, 197, 206, 217, 218,
    220, 224
narratology, 107

objectives, 15, 28, 50, 63, 72, 80, 99,
    105, 122, 157, 191
obstacles, 34, 43, 51
online game, ix, xix, 55, 93, 186–96,
    213, 215, 219, 220
outcome, 7, 10, 11, 20, 29, 99, 100,
    102, 103, 107, 178, 198

perception, 6, 17–19, 35, 36, 43,
    58–60, 62–4, 80, 104, 111, 116,
    125, 166, 167, 177, 178, 224
player/players, x, xiii, 3, 5, 6, 9, 14,
    17–20, 23–31, 35, 36, 38, 39, 41,
    43, 44, 47–55, 58–64, 71–85, 93,
    99–128, 131–6, 138–149, 152,
    155, 157, 158, 160, 163, 164, 166,
    167, 169, 170, 175–8, 182, 187,
    188, 190, 192, 194, 195, 197, 198,
    200–2, 204, 206, 214–16, 218
playful, xi, xviii, 9, 114, 177, 180, 183,
    197, 206, 221
pleasure, xiv, xviii, 22–30, 36, 64, 65,
    104, 110–12, 131, 148, 160, 162,
    164, 199–201, 203, 204, 211, 214,
    216, 219, 220
political correctness, 150, 151, 160,
    219
politicians, xiv, 69, 108, 163

politics, vi, ix, xviii, 16, 100, 120, 129,
    132, 134, 136, 138, 142, 144, 146,
    148, 152–4, 156, 158, 160, 163,
    164, 166, 168, 170, 172, 214, 221,
    223, 224
psychology, ix, xii, 23, 24, 28, 48, 53,
    62, 65, 90, 108–10, 117, 136, 138,
    154, 164, 176, 200, 211, 216, 218,
    223, 224

race, 6, 11, 16, 21, 37, 41, 42, 44, 48,
    61, 73, 77, 88, 90, 96, 105, 120,
    127, 134, 138, 140, 149, 153,
    155–7, 159, 160, 178, 181, 184,
    199, 200, 202, 203, 205, 207, 219,
    225
racism, 32, 134, 203
radio, xiv, 135, 223, 225
real-world, 12, 13, 16, 19, 27, 41, 47,
    49, 58, 64, 93, 100, 101, 104–6,
    167, 168
reality, xiii, 3, 5, 9, 11, 14, 16, 17, 19,
    20, 27, 40–2, 55, 57–60, 62–4, 78,
    86–9, 94, 95, 113–5, 131, 133,
    134, 137–9, 141, 163, 178, 204,
    208, 218, 220
religion, 132, 136, 138, 148, 149, 162,
    221
rewards, 55, 82, 83, 86, 105, 123, 124,
    195
role-playing, xiii, 10, 41, 56, 58, 114,
    120, 140, 152, 153, 157, 187, 190,
    191
rule system, 7, 24, 25, 27, 29, 30
rules, 16, 18, 19, 21, 25–7, 30, 36, 38,
    49, 51–4, 100, 101, 104, 131, 133,
    225

Salen, Katie, 24, 25, 225
sex, 32, 99, 124, 125, 138, 144, 145,
    149–53, 155–60, 166, 199–205,
    212
sexism, 32
shooter, vi, xiv, xviii, 11, 36, 39, 49,
    64, 70–2, 79–83, 86, 120, 132,
    152, 157, 162–72, 215, 222
simulation, x, 14, 16, 17, 64, 81, 83–5,
    88, 92–6, 108, 111, 150, 215, 222,
    224

sound, 17, 19, 26, 33, 34, 38–40, 57, 63, 91, 102, 111, 126, 157, 168, 171

space, v, ix, 7, 11, 15–19, 24, 26, 30, 31, 33–5, 42, 43, 45, 47–55, 57, 58, 60–5, 69, 70, 76, 81, 85, 101, 105, 140–5, 147–9, 159, 163, 165–70, 175–83, 197–200, 206, 209, 213, 215, 218, 220, 222

speed, 5, 8, 9, 57, 59, 60, 65, 89, 138, 189, 209

stereotypes, 99, 116, 133, 134, 143, 145, 152, 156, 160, 201

story, 9–11, 13, 16, 17, 21, 23, 26, 41, 43, 44, 47, 54, 55, 64, 70, 73–6, 79, 82, 86, 87, 90, 91, 94, 104, 120, 121, 133, 139, 144, 148, 167, 168, 179, 186, 190–2, 194, 211–13, 215–17, 219–22, 225

strategy, 13, 24, 35, 39, 63, 113–7, 120, 156, 166, 188, 195

subversive, 79, 160

television, 32, 38, 57, 81, 84, 147, 151, 199, 201, 202, 217, 219, 220, 223, 225

terrorism, 11, 78–83, 85, 116, 146, 147, 169, 222

violence, v, vi, x, xiii, 18, 32, 40, 56, 67, 69–77, 80, 82, 84, 86, 88, 90, 92, 94–6, 99, 108–18, 124, 133, 134, 139, 148–50, 153, 155–9, 170, 172, 176, 178, 182, 211, 212, 215–9, 221–3, 225

virtual, 84, 88, 140, 213, 218–20, 223

war, iii, v, vi, xi, 17, 43, 67, 70, 72, 74, 76, 77, 79, 80, 82–4, 86–96, 134, 170, 188, 208, 209, 213–7, 219, 223–5

weapons, 10, 55, 58, 80, 135, 137, 146, 169, 170, 176

Zimmerman, Eric, 24, 25, 38, 225